The Dread of Plenty:

Agricultural Relief Activities of the Federal Government in the Middle West, 1933-1939

Michael W. Schuyler

Sunflower University Press®

1531 Yuma (Box 1009), Manhattan, Kansas 66502-4228

ISBN 0-89745-117-1

Edited by
Abigail T. Siddall

Layout by
Lori L. Daniel

The Dread of Plenty:

Agricultural Relief Activities of the Federal Government in the Middle West, 1933-1939

To the memory of my father,
Marion F. Schuyler

Glossary

AAA	Agricultural Adjustment Administration
AFBF	American Farm Bureau Federation
CCC	Civilian Conservation Corps
CCC	Commodity Credit Corporation
CWA	Civilian Works Administration
FCA	Farm Credit Administration
FERA	Federal Emergency Relief Administration
FSA	Farm Security Administration
FSCC	Federal Surplus Commodities Corporation
FSRC	Federal Surplus Relief Corporation
Holiday	Farmers' Holiday Association
NRA	National Recovery Administration
PWA	Public Works Administration
RA	Resettlement Administration
REA	Rural Electrification Administration
RFC	Reconstruction Finance Corporation
SCS	Soil Conservation Service
WPA	Works Progress Administration

Contents

Preface

The decade of the 1930s represents a watershed in the history of agriculture in the Middle West. Depression, drought, and savage dust storms combined with rapid political and cultural change to present a challenge of almost indescribable dimensions to the nearly prostrate agricultural community. A more enduring challenge would come, however, from the escalating impact of technology on agricultural production. The increased use of tractors, combines, disk plows, and power drills, together with the introduction of electricity and new hybrid seeds, made life on the farm easier and dramatically increased the farmers' productive capacity. But technological change also resulted in increased capital requirements, led to more ruthless competition, glutted markets, and destabilized farm prices. It did nothing to increase farm markets, to secure a place for small family farmers on the land, or to guarantee that the new abundance would find its way to the poor and needy.

For farmers the grimmest irony of the depression was want in the midst of plenty. The problem of hunger and abundance did not begin, nor would it end, with the depression decade. The collapse of the economy in the 1930s, however, made the poor visible and focused public attention on the irony of poverty and abundance. It was impossible, as the depression deepened and the ranks of the poor increased, not to recognize the economic idiocy, to say nothing of the injustice, of a nation haunted by the dread of plenty.

Paradoxically, while millions of Americans, in both urban and rural areas of the country, were forced to subsist on near-starvation diets, farmers too were impoverished — by abundance. When the depression struck the rural economy with full force in 1931, farm prices, which had been low throughout the 1920s, collapsed as burdensome surpluses glutted the market. By 1932 farm income fell to less than one-half of what it had been only one year earlier. In many areas of the Middle West, land values fell by nearly 70 percent.

Farmers, who emerged from World War I heavily in debt, borrowed even more to expand their production in a futile effort to recover lost profits. They succeeded only in driving prices to still lower levels. As the vicious cycle of overproduction and declining prices continued, a mounting wave of mortgage foreclosures ripped farmers from the land and into the ranks of the urban unemployed.

The farm crisis of the 1930s resulted in a massive intervention by the federal government in the farm economy. New federal agencies, headed by the Agricultural Adjustment Administration, the Farm Credit Administration, the Soil Conservation Service, the Commodity Credit Corporation, and the Resettlement Administration, were created to regulate agricultural production, to raise farm prices, to halt mortgage foreclosures, to stimulate sound land management techniques, and to eliminate rural poverty in the United States.

Unfortunately, in spite of efforts by New Dealers to respond to the needs of drought- and depression-stricken farmers, the problems farmers faced when Roosevelt became President in 1933 remained substantially unchanged on the eve of the Second World War. Indeed, more than fifty years later, the poor and malnourished are still with us, and the farm community is still plagued by overproduction and rapidly changing technologies.

The New Deal did bring about a dramatic change in the relationship between the government and the individual farmer. By the end of the decade, as farm prices remained at disastrously low levels and ruinous surpluses still hovered over the market, farmers had become dependent upon the continuation of the government's massive relief programs for their economic survival. The New Deal's ''emergency'' subsidies had undoubtedly enabled many farmers who would have been forced off the land to continue farming during the depression, but they had not resolved the paradox of want in the midst of plenty nor restored permanent prosperity to the farm economy. The New Deal offered farmers a reprieve, not a solution, to their problems.

Definitions of the Middle West vary widely and have been the subject of endless academic debate. Within the scope of this book the following states will be considered as the Middle West: Ohio, Indiana, Michigan, Illinois, Wisconsin, Minnesota, Iowa, Missouri, North and South Dakota, Nebraska, and Kansas. The diversity of soils and climates in the Middle West enables farmers in the region to engage in virtually every type of agricultural production, but wheat, corn, livestock, and dairy farming structure the agricultural economy of the nation's richest farm belt. In the 1930s, as the leading food-producing region in the nation, the Middle West was hit especially hard by drought, depression, technological change, and overproduction. Farmers in the Middle West were, however, better represented by farm organizations and powerful politicians in Washington than farmers in any other region of the United States. Throughout the decade the Middle West played a primary, and frequently decisive, role in the debate about the future of American agriculture. The government's response to the demands of Middle Western farmers would in large measure define the role of agriculture in American life and thought for the next half-century. This study of agricultural relief activities of the federal government between 1933 and 1939 is intended to provide an introduction to the history of agriculture in the Middle West during the 1930s and to evaluate the impact of the New Deal in shaping the history of the region.

Research for this book was supported in part by the Research Services Council at Kearney State College. I should like to offer a special thanks to the staff of the following libraries and state historical societies for their assistance: Nebraska State Historical Society, State Historical Society of North Dakota, The Land and Its People Museum, State Historical Society of Iowa, Kansas State Historical Society, South Dakota State Historical Society, Franklin D. Roosevelt Library, National Archives, Library of Congress, Herbert Hoover Library, Montana State University Library, University of Oklahoma Library, University of Kansas Library, University of Michigan Library, University of Missouri Library, University of Iowa Library, and the Columbia University Oral History Project. I should also like to thank Donald R. McCoy, my colleagues in the Department of History at Kearney State, and the editors and staff of the Sunflower University Press, especially Robin Higham, Carol A. Williams, and Abigail Siddall, for their support and encouragement. Finally, I want to thank my son and daughter, Adam and Emily, and my wife, Sue, for their patience while I completed this project.

Chapter 1

Hoover, Hyde, Hell and Hard Times — the Republican 4-H Club

The First World War marks an important turning point in the history of agriculture in the United States. During the war, with seemingly unlimited foreign markets and with high prices guaranteed by the government, land values soared to fantastic heights, particularly in the Middle West, as American farmers expanded production to record levels to feed the hungry masses in Europe. After the war, as foreign demand for American agricultural products declined

Small family farmers could not compete with the large, highly mechanized farms of the late 1920s and early 1930s. Here a Middle Western farmer, using nineteenth-century technology, cultivates a corn crop with horses and a hand plow. (Nebraska State Historical Society)

Many farm families in the 1920s and 1930s left the land because they had no choice. For many marginal farmers, however, life on the farm was hard, the work dirty, and the financial rewards minimal. To escape the isolation of rural living, many farmers willingly left to search for a better life in the city. This "model" farmstead captures the isolation and simple lifestyle many wanted to leave behind. (Nebraska State Historical Society)

and the government abandoned its program of artificial price supports, the agricultural boom rapidly turned into depression. There was depression in other industries as well, but the agricultural situation differed in that farm prices fell more rapidly and ultimately fell lower than nonagricultural prices. By rural standards most Middle Western farmers were not poor, but they resented the disparity between agricultural and industrial prices which prevented them from sharing in the highly publicized prosperity of the 1920s.

Farmers and Jeffersonian liberals continued to praise agrarian individualism and self-reliance as the lifeblood from which the nation perpetually renewed its vitality, but by the 1920s agriculture in the Middle West had been transformed from a largely self-sufficing occupation into a highly complex business organized on a mechanized, capitalistic, commercial basis. The farmer was no longer, if he ever had been, an independent yeoman philosophically cultivating the esoteric qualities of rural living, but an entreprenuer in search of profit. Millions of small farms still dotted the landscape, but the inexorable forces of modernization continued to reduce the number of people needed in agriculture. Many farmers left the land willingly. For poor, marginal farmers, life on the land was hard, the work dirty, the hours of labor long and monotonous, and economic rewards minimal. In many rural areas educational opportunities were poor,

Heading wheat on a Middle Western farm. (Nebraska State Historical Society)

Combines and tractors were commonly used by Middle Western farmers in the 1920s and 1930s. The continued mechanization of agriculture resulted in overproduction and a rapid decline in the number of farm laborers. This harvesting scene is from 1933. (Nebraska State Historical Society)

the chance for a full and rich social and intellectual life limited at best. By the late 1920s only one in four Americans were still engaged in farming.

Professing to believe that men raised to industry on the farm were more honest, healthy, moral, and self-reliant than other Americans, agrarian spokesmen at the turn of the twentieth century still relied upon the rhetoric of "rugged individualism" to explain the farmers' contribution to American life and thought. At the same time, however, the farm community increasingly turned to the government for assistance and relief. In the nineteenth century the government responded with legislation, such as the Pre-emption Distribution Act of 1841, the Homestead Act of 1862, and the Timber Culture Act of 1874, to encourage agriculture by providing farmers easy access to cheap land. The Kinkaid Act in 1904 continued the philosophy into the twentieth century. The government also responded, with such legislation as the Morrill Act in 1862, the Hatch Act in 1887, and the Smith-Lever Act in 1917, by developing programs that encouraged farmers to become more scientific, efficient, and productive.

In the 1920s farmers organized to demand that the government do even more to regulate the agricultural economy. In May 1921 a group of Southern and Western Senators, under the leadership of Arthur Capper of Kansas and William Kenyon of Iowa, met in the offices of the American Farm Bureau Federation to organize a "farm bloc" to champion the cause of agriculture in Congress. In the next three years the farm bloc scored legislative successes which, in more prosperous times, would have been considered landmarks in the history of agricultural relief. In 1921 the Packers and Stockyards Act, providing for the inspection and control of rates charged by commission merchants and by stockyards, was enacted into law. In the same year the Grain Futures Act gave considerable power to the Secretary of Agriculture to control speculation on the grain exchanges. In 1922 the Capper-Volstead Act, "the Magna Charta of Co-operative Marketing," exempted farm cooperatives from prosecution under the federal antitrust laws. The Agricultural Credits Act of 1923, which established twelve intermediate credit banks in each of the Federal Reserve Districts, provided loans for those farmers who were not satisfied with the short-term loans from Federal Land Banks. Still, agriculture remained in crisis.

By the mid-1920s farm leaders realized that the most important

Women often helped operate newly purchased "modern" farm machinery. Here three "farmerettes" in Iowa operate a tractor and combine followed by a disc plow. By the late 1920s they were able to cut and thresh 260 acres of wheat in 7½ days. (Nebraska State Historical Society)

Large, highly mechanized farms allowed farmers to plant and harvest more efficiently than ever before. Here three combines, with one box wagon beside each machine, cut a field of wheat or oats. (Nebraska State Historical Society)

problem facing American farmers was surplus agricultural production. In their search for a solution to the problem, farmers were
joined, ironically, by the manufacturers of the farm machinery that
had so greatly increased the farmers' productive capacity. They
hoped to find an answer not by reversing the historic trend toward
more efficiency and productivity, but by developing a complicated
two-price system for marketing agricultural production. George N.
Peek, chief executive officer of the Moline Illinois Plow Company,
and Hugh S. Johnson, a retired army cavalry officer, provided
farmers with a plan designed to give agriculture equality with industry. The price of farm products to be sold domestically was to be
determined by the "fair exchange value" of the crop — the relation
of agricultural prices to purchasing power in the decade before World
War I. Surplus agricultural products were to be purchased at the
domestic price by a federal export corporation and dumped on the
foreign market for whatever price they would bring. If a loss was
suffered by the government, farmers who participated in the program
would be charged an "equalization fee" to compensate the government for its loss. It was assumed that the higher prices received on
farm products consumed domestically would not lead to substantial
increases in production and would more than offset the amount of the
equalization fee. Since grain, meat, and dairy producers sold the
largest percentage of their production at home, the plan had a
tremendous appeal to farmers in the Middle West.

Two important farm spokesmen, Senator Charles L. McNary of
Oregon and Senator Gilbert N. Haugen of Iowa, endorsed the plan
and supported the Peek-Johnson proposals in Congress. With the
support of the powerful American Farm Bureau Federation, the
McNary-Haugen legislation became a stormy political issue
throughout the decade of the 1920s. In 1927 and again in 1928 bills
passed Congress only to meet with vetoes from President Calvin
Coolidge. The President, who was strongly influenced on agricultural matters by Secretary of Commerce Herbert Hoover, objec-
ted that the bills asked for an improper delegation of the taxing
power, would result in price-fixing, and would create a cumbersome
bureaucracy. Coolidge claimed that the bills were economically
unsound, that the higher prices contemplated by the bills would lead
to further overproduction and larger surpluses, and that the disposal
of American goods abroad by dumping would arouse foreign resent-

This cheap house, surrounded by expensive machinery, shows the heavy investment required to be an efficient farmer by the late 1920s and early 1930s. (Nebraska State Historical Society)

ment and promote retaliation. Coolidge's decision to veto the McNary-Haugen bills was sound. Not only would the plan have been difficult to administer, the increase in farm prices undoubtedly would have resulted in increased production and steadily mounting surpluses. Foreign nations had already erected high tariff walls and were prepared to adopt antidumping laws if the McNary-Haugen plan were put into effect. The problem was not that Coolidge rejected the McNary-Haugen proposals, but that he offered no substantive alternative to take their place.

The American Farm Bureau Federation continued to champion McNary-Haugenism throughout the 1920s. The oldest of the farm organizations, the National Grange, endorsed an alternative plan for agricultural relief, the so-called export-debenture plan. According to the proposal, which was worked out by Prof. Charles L. Stewart of the University of Illinois, a bounty, amounting to one-half of the existing tariff rate, would be paid by the government to exporters of American farm products. Advocates of the plan anticipated that farmers would benefit not only by receiving a higher price for exportable farm commodities, but also, because of the reduced domestic supply, from higher domestic prices. Like the Peek-Johnson proposals, the export-debenture plan assumed that dumping American farm products abroad would remain feasible indefinitely and failed to call for production controls.

The Farmers' Union, the most militant of the farm organizations, demanded that the government endorse yet another plan for agricultural relief. The Farmers' Union wanted officials in Washington to fix prices, as they had during the war, based upon the farmers' cost of production plus a guaranteed profit. After the cost of production had been determined for individual crops, processors would be required to purchase farm products consumed domestically at that price. Surplus farm products would be stored on the farm or dumped on the foreign market for whatever price they would bring.

The cost-of-production proposal was considered more radical than the other plans, but the thinking of the farm organizations was quite similar. Each plan aimed at restoring the balance between agricultural and nonagricultural prices that had existed before the war; each called for the government to play a more active role in the economy. None of the plans contemplated acreage reduction or production controls; none of the farm organizations sought a solution

World War I and the early 1920s witnessed the expansion of acreage under cultivation, frequently into marginal areas of the Western and Northern Plains. Here a South Dakota farmer works the land on the treeless plains. (Nebraska State Historical Society)

In spite of the rapid mechanization of agriculture, most Middle Western farmers still used horses and mules to work their farms in the 1930s. (The Land and Its People Museum, Red Cloud, Nebraska)

to the farm crisis by redistributing surplus agricultural production to the poor. The farm organizations continued to look abroad, not to changes at home, for a solution to the problems of the American farmer.

Coolidge's successor, Herbert Hoover, was convinced that farmers could solve their problems by organizing along the same lines as the business community. Like his predecessors, President Hoover argued that farmers should become more rational and efficient in their production and marketing practices. He also recognized the need, however, for limiting agricultural production. Still, Hoover was unwilling to tamper with the delicate mechanism of "American individualism" and placed his faith in voluntary cooperation to restore balance to the agricultural economy.

On June 29, 1929, Hoover's ideas were endorsed by the Congress when it passed the Agricultural Marketing Act and set up a new Agency, the Federal Farm Board. The primary function of the Farm Board was to lend money to cooperatives and to regulate the movement of farm commodities to market by improving marketing facilities. A peripheral part of the Board's program called for the creation of "stabilization" corporations, if the need arose, to purchase ag-

Milo Reno, head of the militant Farmers' Holiday Association. Reno led a bitter attack against President Hoover, and later against President Roosevelt and Secretary of Agriculture Henry A. Wallace, in the early 1920s. (State Historical Society of Iowa)

ricultural surpluses and to hold them from the market until the demand for farm products rose. The President was willing to assist farmers in building a new cooperative structure that would enable them to control their own production. He was unalterably opposed to the government's dictating mandatory production controls.

Any chance that Hoover's plan would work ended when the depression struck the farm economy. Faced with sagging farm prices, the Farm Board began buying wheat and cotton in a vain attempt to stabilize agricultural prices by withholding surpluses from the already glutted farm market. Unfortunately for Hoover, the activities of the Farm Board's stabilization corporations were intended to solve only minor price variations and were not able to handle the mounting surpluses. When it became apparent that the Farm Board would not be able to purchase enough farm products to keep agricultural prices from falling to desperately low levels, the Farm Board ceased its buying activities. Hoover and his Secretary of Agriculture, Arthur Hyde, again pleaded with farmers to voluntarily cut back their production. The President, fearing that the expansion of federal power would undermine rural values, still refused to call for mandatory production controls. By the end of his administration Hoover's commitment to values of "American individualism" led to increasing resentment, exasperation, and, ultimately, chaos. The price of farm products continued to spiral downward.

Farm income in 1932 had fallen to $4,377,000,000, a drop of 58 percent from the income farmers had received in 1929. The value of farm exports, which had totaled $1,821,000,000 in 1929, had fallen to only $590,000,000 in 1932. While the general price index for nonagricultural prices fell by 30 percent during the depression, the price of farm products fell an average of 56 percent, adding even further to the disparity between agricultural and nonagricultural prices. Farmers in the Middle West were in an especially disadvantageous position, since the price of grains and meat animals, which structured the economy of the farm community, fell even more rapidly than other agricultural prices during the depression. Farmers who had mortgaged their farms to expand production during the war years faced a mounting wave of foreclosures. The Bureau of Agricultural Economics estimated that for the five-year period ending March 2, 1932, 20 percent of the farms in the United States had changed hands through forced sales, and another 3.5 percent had been sold because of tax delinquency. By the end of 1932 there were more forced than voluntary sales of farms.

In the summer of 1932 a new farm organization, which would translate the confusion and frustration of Middle Western farmers into a program of militant activism, was established in Iowa under the leadership of the former president of the Iowa Farmers' Union, Milo Reno. Reno, who was sixty-five years of age in 1932, had fought farmers' battles with unrelenting energy since the turn of the century. A former Greenbacker and Populist, Reno's colorful rhetoric and bombastic diatribes against the Hoover Administration brought back memories of the agrarian upheavals which had inflamed the Middle West since the nineteenth century. By late 1932 Reno had decided that just as "it is impossible to get pure water from a sewer," it was impossible to expect anything of "virtue" from the Hoover admin-

The dramatic increase in farm foreclosures resulted in a "penny auction" rebellion in the Middle West. Here farmers near Elgin, Nebraska, supported by the Farmers' Holiday, try to save a neighbor's farm. (Nebraska State Historical Society)

istration. With a program demanding inflation, mortgage refinancing, and cost-of-production, the Farmers' Holiday Association
attracted a small but aggressive following in the Middle West. Each
of the Farmers' Holiday Association's demands was considered
"radical" by Reno and his followers and provided a common ground
of agreement for the agrarian left in the early 1930s.

Reno believed that inflation would not only benefit poor farmers
by reducing debts and raising farm prices, but also destroy the money
power of the East, which, in his mind, dominated and exploited the
American farmer. The cost-of-production plan was also considered
to be much more than a proposal to raise agricultural prices. It
represented, for many agrarian radicals, a symbolic attack on the
established power structure in America. Reno was convinced that the
government, as well as the financial system, was controlled and
manipulated by the wealth and power of the nation's business elite. If
the government could be persuaded to fix agricultural prices, that
would be direct evidence, not only that farmers were to receive their
fair share of the nation's wealth, but also that the control of the
government had been restored to the people and that the money-
changers had been driven from the temple. The demand that farm
mortgages be refinanced was radical not only because it called for an
expansion of federal power to meet the needs of agriculture, but also
because it seemed to challenge the sanctity of contracts and promised
to replace the "Usurers" of the East with a more responsive and
paternalistic government. The most "radical" feature of Reno's
appeal was not his demand for inflation, cost-of-production, or
mortgage refinancing, but his reliance upon a new technique, the
farm strike, to articulate the demands of the Farmers' Holiday
Association.

As early as 1927 Reno had suggested that if legislators failed to
meet the farmers' demands, the farm community should go on strike.
While some of Reno's followers apparently believed that, if farmers
withheld their products from market, prices would rise, the farm
strike, without provisions for controlling or distributing surpluses,
made little economic sense. Even if farmers withheld their products
from market, the eventual release of the surpluses stored on the farm
would drive prices down again. The Farmers' Holiday Association,
like the American Farm Bureau Federation, the Grange, and the
Farmers' Union, simply failed to come to grips with the problem of

overproduction. The farm strike was, however, an effective instrument to dramatize the farmers' dilemma and to bring pressure on state and federal officials to do something about the agricultural situation.

In August 1932, as the farmers' mood turned from quiet desperation to angry frustration, Milo Reno launched a farm strike in the Middle West. Farmers vented their rage against the established order by parading the streets of Middle Western cities with signs reading "Hoover, Hyde, Hell and Hard Times — The Republican 4-H Club," and by threatening to halt mortgage foreclosures. When the movement spread throughout Iowa and into a number of neighboring states, farmers blocked roads, dumped milk, and spread nails on the highways to prevent agricultural supplies from flowing to the cities. As the strike gained momentum and violence flared in the farm belt, Reno warned that a revolution was underway and that there was nothing he could do to halt the farmers' drive for relief. The problem of plenty was getting serious.

Franklin D. Roosevelt delivering a major farm speech at the state capitol in Topeka, Kansas, during the 1932 campaign. (Kansas State Historical Society)

Chapter 2

The New Deal Begins

Genesis of the Farm Program

With the nation on the verge of economic collapse, and with dire predictions that revolution was just around the corner, Franklin D. Roosevelt launched his campaign to win the Presidency. Roosevelt, in spite of his aristocratic background, viewed himself as a rural agrarian democrat and instinctively identified with the Middle West and the rural way of life.

A key element in Roosevelt's strategy was to attract the Middle Western farm vote. When the Republicans again nominated Hoover, many farm spokesmen openly expressed their hostility and disillusionment. William Lemke, a Republican candidate for Congress from North Dakota, spoke for many traditional Republican farmers in the Middle West when he wrote to Senator George Norris of Nebraska, ''I feel first that Hoover is not a Republican and secondly that in this crisis one should arise to the highest of American citizenship first and party lines afterward. I do not believe this nation can endure four more years of Herbert Clark Hoover.'' As governor of New York Roosevelt had supported conservation and land-use planning but was vague about how he would restore prosperity to agriculture. With consummate political skill Roosevelt promised something to everybody while continuing a dialogue with proponents of cost-of-production, the export-debenture plan, and the McNary-Haugen bills. Publicly, during the early stages of the campaign, Roosevelt continued to walk a tightrope of noncommitment, but behind the scenes Rexford Tugwell, one of Roosevelt's agricultural advisors, discovered a new plan for farm relief that would ultimately provide the framework for the New Deal's farm program — the voluntary domestic allotment plan.

The domestic allotment program called for farmers to enter into a contract with the government to limit production by reducing their acreage to an amount determined by agricultural officials in Washington. Farmers in turn were to be given benefit payments, financed from taxes collected from processors of agricultural products. The plan, which had been developed by W. J. Spillman, John D. Black, Beardsley Ruml, and especially M. L. Wilson, was to be decentralized and would be based upon voluntary cooperation, not compulsion. Unlike other farm proposals being debated in the early 1930s, the domestic allotment plan, by calling for controlled production, emphasized the problem of overproduction while recognizing the farmers' changing position in the world of international trade. It was assumed, given the financial incentives offered by the government, that enough farmers would voluntarily participate to guarantee the success of the program.

Walter Lippmann would later call the plan "the most daring economic experiment ever seriously proposed in the United States," but it was a logical outgrowth of previous efforts to apply modern principles of scientific planning and management to the farm economy. The proposal came not from farmers, but from agricultural economists and academics trained in various social science disciplines. It was not designed to attack the problems of land abuse, farm credit, rural poverty, or the maldistribution of wealth. Wilson, who conceded that the plan was the "most conservative" of the various farm plans being debated during the campaign, was interested in other democratic reform issues but concluded that the government's first task was to raise agricultural prices by controlling burdensome farm surpluses.

The domestic allotment plan was generally unpopular with the nation's poorest and most radical farmers. They understood that although they would benefit from higher farm prices, their problem was not so much that they were plagued with unmarketable surpluses but that they were not able to produce enough to escape their marginal existence. Government payments to cut back their production would do little to improve their economic status or to change their relationship with the land. By contrast, wealthier farmers, if they retired large sections of land from production, would benefit both from higher prices and from substantial financial payments from the government. When Henry I. Harriman, the president of the United States Chamber of Commerce, openly endorsed the domestic allot-

An important part of Roosevelt's campaign strategy in 1932 was to win the Middle Western farm vote. Here, the Democratic candidate in the wheat belt near Colby, Kansas. (Kansas State Historical Society)

ment proposal, a number of agrarian radicals, who were already suspicious of the plan, became convinced that Wilson's plan was but a ''gesture to mislead.'' Although, as Rexford Tugwell later argued, the domestic allotment plan should not be criticized too harshly for what it did not propose to do, the conservative thrust of the proposal would have an important and lasting effect on the New Deal's farm program and the future of agriculture in the United States.

Wilson, along with Rexford Tugwell, tried to persuade Roosevelt to endorse the domestic allotment idea during the Presidential campaign. Roosevelt appeared to accept the concept but would not openly endorse the domestic allotment plan for political reasons. Important farm leaders, like William Lemke of North Dakota, who wanted the government to refinance all farm mortgage indebtedness by issuing federal greenbacks, William A. Hirth, president of the Missouri Farmers' Association, Senator George Norris, chairman of the National Progressive League for Roosevelt, Edward A. O'Neal, president of the American Farm Bureau Federation, John A. Simpson, president of the Farmers' Union, and George N. Peek, the creator of the McNary-Haugen proposals, remained convinced that Roosevelt was supporting their individual plans for agricultural relief and actively supported him during the campaign.

Roosevelt's major farm speech during the campaign was delivered in Topeka, Kansas, in September 1932. His choice of Topeka as a location for his first major agricultural speech was part of a carefully conceived strategy to capture the Middle Western farm vote. Roosevelt wrote to Farmer-Labor Governor Floyd B. Olson of Minnesota that the chief thought he wanted to get across was that ''I have a deep and very practical and personal interest in the farm problem and that I am willing to try things out until we get something that works.'' At least twenty-five people either submitted memoranda or read drafts of the speech, but the heart of Roosevelt's Topeka address was taken from a manuscript drafted by M. L. Wilson.

Roosevelt talked about a number of farm problems, such as rural taxation, the farm debt, and the planned use of the land, but the major thrust of the Topeka speech was concerned with proposals to control surplus agricultural production. He emphasized that a satisfactory farm program should insure agriculture tariff equality with industry, that it should involve no export dumping, that it should be voluntary and decentralized, and that it should be self-financing. While Tugwell and Wilson worried that Roosevelt might be wavering in his support for the domestic allotment plan, the speech, by not mentioning production controls, processing taxes, or the possibility of reduced acreage, had offended no one and enabled him to keep his support from the major farm organizations intact. Raymond Moley, a member of Roosevelt's Brain Trust, later concluded, ''More than any other single speech — in the entire campaign — it captured the votes of Middle Western farmers. . . . It won the Midwest without waking up the dogs in the East.''

In additional speeches in Sioux City, Iowa, Springfield, Illinois, and Atlanta, Georgia, Roosevelt emphasized the interdependence of the agricultural and industrial sectors of the economy and repeated the points outlined in his Topeka address. His campaign was running smoothly, but John A. Simpson and Milo Reno were increasingly impatient with Roosevelt for failing to openly endorse their cost-of-production proposal. In Iowa, Henry A. Wallace, the influential editor of *Wallace's Farmer and Iowa Homestead*, warned Roosevelt to keep his eye on discontented farmers in the Middle West. Wallace wrote to Roosevelt just before his Sioux City speech: ''In my opinion, the economic abuses from which they have suffered have been more severe than those from which the colonists suffered during the period from 1763 to 1775. The Holiday movement may well prove to be a Boston Tea Party if we do not get rid of the Lord North's and George III's.''

During the fall of 1932 President Hoover began his campaign for re-election in the heartland of rural militancy, his home state of Iowa. While angry farmers paraded with placards denouncing Hoover's farm program, the embattled President vigorously defended the record of his Administration on agricultural relief. Hoover promised farmers that if reelected he would do something about mortgage foreclosures, would improve credit facilities, would ask for the repeal of the stabilization features of the Agricultural Marketing Act, would implement a sound land-use planning program, and would work for the improvement of the farmers' trading position in world markets. Throughout his campaign Hoover warned that government efforts to control production would result in regimentation, coercion, and autocratic government. He observed in his memoirs that government production controls, ''. . . despite all the clatter about 'voluntary agreements' and 'democratic action,' were as wide a departure from a society of American free men as those of the fascist regimes of Italy and Germany. And they were wholly unnecessary and ineffectual as a remedy of our farm problems.''

In spite of Hoover's assurances, as William Allen White later observed, people were ''scared into a blue funk'' by the collapse of the economy. Hoover's only chance of winning the farm vote was to convince farmers that Roosevelt would cause an even greater disaster. Roosevelt appeared to be most vulnerable on the tariff issue. Along with other Democrats, he had consistently criticized the high tariff policies of the Republicans and blamed the Hawley-Smoot tariff, which had become law in 1930, for the decline in agricultural exports. Hoover, who had witnessed a decade of agitation by farmers to maintain an effective tariff to protect American farm markets from foreign competition, warned farmers that ''the removal of or reduction of the tariff on farm products means a flood of them into the United States from every direction, and either you would be forced to still further reduce your prices, or your products would rot on your farms.'' Roosevelt immediately clarified his statements on tariff reform. He assured farmers, ''I do not intend that such duties shall be lowered. To do so would be inconsistent with my entire farm program, and every farmer knows it and will not be deceived.''

Franklin D. Roosevelt, with Kansas Governor Harry Woodring at the microphone, campaigning for the farm vote in 1932. Roosevelt remained vague about his farm program but swept the Middle West in November. (Kansas State Historical Society)

Hoover continued to warn against Roosevelt, but his efforts on behalf of agriculture were all but forgotten as the nation plunged even deeper into depression. Rexford Tugwell observed in his diary that the positive programs of the Hoover presidency were "drowned in the advancing tide of disaster, pitiful dissolving monuments to the Great Engineer in the White House. He was a figure of tragedy by now: A Canute. His dignity was lost in a spate of ridicule. . . . Louis Howe was heard to say that he could be beaten by a Chinaman in the approaching election."

Roosevelt swept the Middle West, winning every state, enroute to a smashing victory in November. Ironically, radical and moderate farm leaders had joined together to work for Roosevelt's election, but the President-elect had won without gaining the support of the major farm organizations for the domestic allotment plan. W. R. Ronald, the editor of the Mitchell, South Dakota, *Evening Republican*, warned Roosevelt not to be fooled by the support he had won in the farm belt. He pointed out that there was still a wide diversity of opinion in and out of Congress on the question of farm relief. While Ronald suggested that Roosevelt call a conference of farm leaders to discuss agricultural relief, he correctly prophesied that it would be necessary for Roosevelt to take the lead if there was to be any hope of reaching an agreement on a satisfactory farm bill. Other farm leaders, like William Hirth of Missouri, also expressed concern that Roosevelt had surrounded himself with ''professors and hair-splitters'' who did not understand the gravity of the farm crisis. Hirth warned that ''a sound agriculture is the only thing that stands between us and revolution — we may avoid the latter for a few years through the adoption of some temporary palliative but in the end it will come, and the French Revolution will have been a mere side show in comparison.''

In September 1932 Milo Reno had called an end to the farm strike in the Middle West, but by January 1933 farmers in the region again acted out their frustrations by attacking the tangible economic forces that threatened their existence. While Congress vacillated during Hoover's final days in office, Edward O'Neal and John Simpson warned that a revolution was about to sweep the countryside. The Farmers' National Relief Conference, which included a number of Communists and various groups which supported the Farmers' Holiday Association, urged farmers to denounce the theory of surplus production and demanded that the government buy farm surpluses to give to the starving masses in the cities. The questions of whether the farm organizations should align themselves with the urban poor, and whether they should resort to violence and coercion, increasingly split the farm organizations into opposing camps.

Government inaction, combined with escalating mortgage foreclosures, finally convinced many farmers that it was necessary to take the law into their own hands. They refused to see the loss of their farms as simply another wave of business failures; they were being driven off the land and from their homes. With violence spreading through Iowa, Wisconsin, Minnesota, Nebraska, and the Dakotas, militant farm leaders grew bolder in their demands, particularly on the state and local level. On January 14, 1933, officials of the Farmers' Holiday Association in Sidney, Nebraska, warned that two hundred thousand members of the Holiday would gather in Lincoln on February 15 to tear down the capitol building if the state legislature did not enact a law proclaiming a moratorium on mortgage foreclosures. By January 21 the state legislatures of Iowa, Minnesota, Wisconsin, Illinois, Indiana, Ohio, Michigan, and Kansas had received similar requests for moratorium legislation.

On February 16 five thousand farmers, led by the president of the Indiana Farm Bureau, William H. Settle, marched on Indianapolis to protest the failure of the state legislature to pass tax-relief measures that farmers had previously demanded. Settle urged farmers to support a tax strike unless relief was immediately forthcoming. In Nebraska four thousand farmers marched to the capitol in Lincoln to reinforce their demands for mortgage relief. H. C. Parmenter, president of the Nebraska Farmers' Holiday Association, explained that the farmers were in revolt against "international bankers." He demanded that the federal government issue greenbacks to repay farmers for all losses incurred in recent years as the result of bank failures and farm debts, reduce government salaries by 25 to 50 percent, cease mortgage foreclosures, and provide longterm federal refinancing of all farm debts at 3.5 percent interest.

Farmers soon became more resourceful in efforts to save their farms. Rather than halt foreclosure sales, they allowed them to proceed but controlled the bidding. After purchasing a farm and its implements at ridiculously low prices, they returned the property, with a free title, to the original owner. If the mortgage-holders tried to stop the sale, the sullen farmers threatened retribution. As the "penny auction" rebellion spread throughout the Middle West, life insurance companies, which held a large percentage of farm mortgage indebtedness, became more hesitant to foreclose on farm mortgages. There was also the promise of action in Washington.

With endorsements from the American Farm Bureau Federation, the Farmers' National Grain Corporation, the National Livestock Marketing Association, the National Grange, and the Farmers' Union, a joint resolution was introduced in Congress asking that state courts hold up foreclosure proceedings until Congress had time to act. Although Congress did nothing as the moribund machinery of the government ground to a near halt, state governments were more responsive to the political pressure of the farm groups. In Iowa, Indiana, Minnesota, Nebraska, North Dakota, South Dakota, and Wisconsin farmers were granted moratoriums, either by executive request or by legislative action, on farm foreclosures. But in spite of the receptiveness of state officials, the economic position of the Middle Western farmer continued to worsen.

As the depression deepened, the rhetoric of violence and revolution escalated. Senator Henrik Shipstead, a Farmer Laborite from

Minnesota, in an address before the League of Industrial Democrats in Washington, warned, "We are in the midst of an economic revolution right now in this country, as can be shown by the happenings every day in all parts of the nation. . . ." He castigated capitalists who had "taken so much out of the farmer that they have killed the goose that laid the golden egg." A South Dakota farmer summarized the feelings of many Middle Western farmers when he wrote, "It seems like the people are not going to stand for this situation much longer. It will not surprise me that revolution may break out at any time which I hope will never happen."

The Communist Party, which had been active in Middle West since the 1920s, was optimistic that the revolution was about to begin. Organizers like Mother Ella Reeve Bloor, Harry Lux, and Harold Ware worked to infiltrate the Farmers' Union and the Holiday Association. Working through the Farmers' National Committee for Action and the United Farmers' League, Communists in the Middle West gained some influence and enjoyed the respect of large numbers of farmers, particularly in Nebraska and the Dakotas, in the early 1930s. They were popular, however, not because farmers looked to collectivization as a solution to the farm crisis, but because the Communists promised to attack the money power in the United States and joined militant farmers in demanding government relief for agriculture.

Farmers were still largely unorganized, but the materials for revolt, low prices and continuing farm mortgage foreclosures, were seen as real threats to the established political order in the Middle West. On the eve of Roosevelt's inauguration, leaders at the Farmers' Holiday Convention in Bismarck, North Dakota, urged farmers to organize county defense councils

> ". . . to prevent foreclosures, and any attempt to dispossess those against whom foreclosures are pending if started; and to retire to our farms, and there barricade ourselves to see the battle through until we either receive cost of production or relief from the unfair and unjust conditions existing at present; and we hereby state our intention to pay no existing debts, except for taxes and the necessities of life, unless satisfactory reductions in accordance with prevailing farm prices are made on such debts."

Milo Reno, who had decided that it was time to stop "pussy-

Henry A. Wallace, Roosevelt's Secretary of Agriculture, meeting with the Bremer County Farm Bureau, Waverly, Iowa. (Franklin D. Roosevelt Library)

footing around," warned that if Roosevelt did not redeem his pledges to agriculture, ". . . there will be a nation-wide strike called and we of the middle west, at least, are ready for the contest."

A Reaper Or a Ferris Wheel?

An atmosphere of crisis cast an ominous shadow over the White House when Franklin D. Roosevelt took office on March 4, 1933. The President moved quickly to name Henry A. Wallace to head the Department of Agriculture. Wallace, whose father had been Secretary of Agriculture under Harding and Coolidge, was a natural choice for the appointment. Tired of the "hypocritical pretense of trying to be a progressive and a Republican at the same time," Wallace deserted the Republican Party to support Roosevelt during the campaign and soon became one of Roosevelt's most articulate supporters in the corn belt. The young Iowan's rhetoric often sounded like that of his populist forebears, but he occupied a middle ground during the depression and would serve as a major force of moderation in the farm belt.

Before he became Secretary, Wallace had suggested that a "peaceful revolution'' might be necessary and speculated that a kind of ''Christian communism'' might be required to eliminate the suffering that had resulted from the capitalist system. He retained, however, a commitment to the private ownership of property and to the traditional market place economy. Wallace agreed with Roosevelt that long-range planning was necessary for agriculture, but he was reluctant to support production controls. In April 1932 he warned Roosevelt that the domestic allotment plan would lead ''toward a bureaucratic plan of state socialism,'' but by late summer he had decided it was the only feasible plan for farm relief. Wallace had no illusions about the task he would face in Washington. After he accepted the nomination he wrote in his farm journal, ''It is fairly easy to put out a fire before it gets much of a start. To put it out after wind and time and neglect have fanned it into a flaming rage is a task of greater difficulty. The new administration must make up for twelve years of lost time.''

Wallace was also aware that this selection was viewed with suspicion by many farm leaders. George N. Peek, whom Wallace had supported for the post, and cost-of-production advocates, like William Hirth, complained that Wallace was a ''hairsplitter'' who didn't have enough ''hair on his chest.'' Milo Reno, who made no effort to conceal his displeasure with the appointment, doubted that the ''erratic and academic'' Wallace ''would be any improvement over the present jackass who occupies that position.''

Roosevelt was determined to build a consensus among the major farm organizations for a new farm program. At his direction, Wallace called fifty farm leaders to Washington on March 10 to discuss agricultural relief. The conference resurrected nearly every relief measure that had been proposed during the past half-century. The American Farm Bureau Federation fronted for the domestic allotment plan, but many of its leaders continued to support the export dumping schemes formulated by George Peek and Hugh Johnson in the 1920s. The Farmers' Union and the Farmers' Holiday Association agitated for cost-of-production, while the Grange again endorsed the export-debenture plan. Unable to agree among themselves, the farm leaders ultimately recommended that Roosevelt should have absolute power to fix prices, to regulate the marketing and processing of agricultural products, and to lease land or use some other means of curtailing production. Reeling from the effects of the depression, farm leaders attending the conference offered Roosevelt complete dictatorial powers over American agriculture.

After his inauguration, Roosevelt called a special session of Congress to deal with the banking emergency but did not plan to deal with the agricultural crisis until the regular session of Congress. At the urging of Wallace and Tugwell, however, the President decided to ask for farm-relief legislation during the special session. While Roosevelt's advisors worked on a new farm bill, the Farmers' Union and the Farmers' Holiday Association issued a call for legislation far beyond the scope of Roosevelt's still unannounced plans for agriculture.

The Farmers' Union, which was meeting in executive session in Omaha while the national agricultural conference was meeting in Washington, commended Roosevelt for his "courageous stand" and placed itself on record as agreeing with his program, which, the Union apparently believed, included cost-of-production. At the same time, however, the executive session adopted resolutions supporting the Frazier-Lemke bill, which proposed that the government re-finance all farm indebtedness by issuing federal greenbacks, urged the remonetization of silver, and advocated drastic freight-rate revi-sions. The Farmers' Union had been represented at the national agricultural conference by W. P. Lambertson, but John A. Simpson, the president of the organization, took no part in the conference. Simpson had met with Roosevelt on March 6 and was informed that nothing would be done concerning agriculture during the special session. When Roosevelt changed his mind, Simpson was in the West and could not be reached. The failure to contact Simpson was to cause grave difficulties for the Administration in the future.

The Farmers' Holiday Association, headed by Milo Reno, was also meeting in Des Moines, Iowa, when Wallace called the national farm conference. Enthusiastic representatives from sixteen states, including eleven states in the Middle West, listed their legislative demands and threatened a national farm strike if the objectives of their program were not met by May 3. Among the demands of the Holiday were passage of the Frazier-Lemke bill, an honest dollar, federal operation of all banks and credit institutions as public utilities, cost-of-production for agricultural products, payment of the soldier's bonus, and a steeply graduated income, gift, and inheritance tax at near-confiscatory rates.

Striking farmers, suffering from low prices and glutted markets, block highways to prevent farm products from going to market. (State Historical Society of Iowa)

The Farmers' Holiday also came out strongly against the domestic allotment plan. Reno was convinced that the proposal represented "the same old tactics to hand the people a little measure of relief to suppress rebellion, with no intention of correcting a system that is fundamentally wrong." R. A. Wright, of Neosho Falls, Kansas, who served as the chairman of the legislative resolutions committee, announced, "We shall regard a gesture of this sort in the present crisis as worse than silly and denounce in no uncertain terms any effort that has been made or may be made to revive this measure as a deliberate effort to palliate rather than remedy our agricultural ills." As Reno and Simpson joined forces to demand inflation and cost-of-production, leaders of the Farmers' Holiday Association gravely informed Roosevelt, "We do not desire to seek redress of our wrongs and grievances through force except as a last resort, but we are free men and we refuse to become serfs and slaves of the usurer and money king." Reno, who argued that the Farmers' Holiday Association had accomplished more in ten months than the other farm organizations had accomplished in their entire history, was convinced that he could force Roosevelt to endorse the Holiday Association's demands for rural relief. He concluded that the Holiday movement had "aroused the moral conscience of all groups of society as to the unthinkable condition of not only the American farmer, but the starvation, privation and want in a land of super-abundance." He promised not to unleash an agrarian revolution in the countryside until Roosevelt had a chance to act on his demands. Before the end of the year, however, Reno and his followers would feel compelled to rely upon their "last resort" in an effort to force Roosevelt to recognize the Holiday Association's legislative program.

On March 16 Roosevelt sent a special message to Congress asking for the passage of a farm bill, entitled the Agricultural Adjustment Act, to bring relief to the agricultural sector of the economy. The aim of the bill was clear — to restore farm purchasing power to a position comparable to the equality agriculture had enjoyed with industry in the five-year period from 1909 to 1914. To realize that goal, the bill included provisions to implement virtually every farm-relief measure, with the exception of the cost-of-production plan, which had been suggested since the 1920s. The Secretary of Agriculture was given "discretionary" powers to (1) grant farmers benefit payments for reducing acreage in production, (2) form marketing agreements with processors, farmers' associations, and other groups that handled farm products, to promote higher prices and to eliminate surpluses, (3) levy processing taxes to finance the program, and (4) use revenues gained from processors to subsidize losses incurred on farm exports. It was obvious that the Administration, which had drafted the bill before the farm leaders were called to Washington, had decided to ask for broad legislation which would attract the support of as many farm organizations as possible. Tugwell pointed out, "What we were looking for was a way of controlling production which would be politically feasible." The approach may have been necessary to gain quick passage of the bill but, as George Peek concluded, "The Farm organizations had very little to do with the bill, and it involved some new principles which they knew nothing about."

When debate on the bill began, the lines of support and opposition were quickly drawn. Edward O'Neal, president of the American Farm Bureau Federation, C. E. Huff of the Farmers' National Grain Corporation, C. G. Henry of the American Cooperative Association, Louis J. Taber of the National Grange, Charles A. Ewing of the National Livestock Marketing Association, and Ralph Snyder of the

National Committee of Farm Organizations, sent letters to each Congressman, supporting the measure and urging its enactment. Efforts to build support for the farm bill met with only moderate success.

During the Congressional debate the Kansas City Chamber of Commerce sent questionnaires to six thousand "dirt farmers" in 484 counties in Kansas, Missouri, Nebraska, Colorado, and Texas to determine their attitudes about farm relief. Farmers expressed an overwhelming sentiment for federal refinancing of mortgage indebtedness, voted four to one against Hoover's Federal Farm Board, were three to two against the domestic allotment plan, and two to one against any kind of crop controls. Farmers wanted government help, but many still believed that the the "natural laws" of economics would resolve the problem of overproduction. Dan Casement, who would speak for conservative farmers throughout the 1930s, wrote to William Allen White that he was against the bill "because I can't for the life of me see how it can fail to play merry hell not only with us farmers but with the whole damn country. . . . And I'll be damned if I want a scut like Hank Wallace messing in my business."

On March 21 the House voted to consider Roosevelt's farm bill, without amendments and with debate limited to four hours. Opponents of the bill charged that it did not have the support of farmers, would create an army of tax-gatherers, would threaten the farmers' individualism by creating a cumbersome bureaucracy, and would lead the United States down the path of Communism. At the same time a constant theme of "follow the President" also permeated the debate in the House. Roosevelt's popularity, combined with an atmosphere of crisis and the shrewd parliamentary move to limit debate to only four hours, was enough to insure the bill's passage in the House. On March 22 the bill passed by a large majority, 315 to 98. Of the 137 Representatives from the Middle West, only 26, including 12 Democrats, 13 Republicans, and 1 Farmer-Laborite voted against the bill.

Supporters of the Farmers' Holiday Association and the Farmers' Union immediately protested the House version of the farm bill because it did not provide for cost-of-production. Fred Schultheiss, an executive member of the Farmers' Union, advised Roosevelt to listen to Simpson and Reno, not to farm leaders of the "selfstyled and silkshirt variety, the taxeaters and parasites who have been nursing on fat salaries connected with the relief measures as passed in the last twelve years.'' Simpson also wrote Roosevelt that the only valid test of any legislation passed by Congress was whether it put money into the hands of the poor, and warned, ''Unless the legislation being passed by Congress results in doing that, God pity those who are responsible for passing the legislation.''

Another frequent complaint from the Middle West was that the farm bill did nothing to improve the credit position of the farmer or stop the disastrous increase in farm foreclosures. On March 20 Representative William Lemke wrote Roosevelt asking him to throw his support behind the Frazier-Lemke bill to refinance farm mortgages. Lemke pointed out that seventeen state legislatures, including those in North Dakota, South Dakota, Nebraska, Iowa, Minnesota, Wisconsin, Illinois, and Indiana, had memorialized Congress to pass his bill. Roosevelt turned a deaf ear to Lemke's demands, but the Administration was already working on a plan to improve the agricultural credit situation. On March 27, by executive order, Roosevelt placed federal credit activities, which had been under the supervision of the Secretary of Agriculture, the Reconstruction Finance Corporation, and the Federal Farm Board, under the direction of a new agency, the Farm Credit Administration.

By early April a bill entitled the Emergency Farm Mortgage Act, which had been drafted under the supervision of Henry Morgenthau, Jr., and William I. Myers, of Cornell, was ready to be submitted to Congress. On April 5 Roosevelt sent a message to Congress outlining the major objectives of the bill and urging its quick passage. The provisions of the bill were designed to (1) establish a uniform rate of interest on all federal farm loans, (2) refinance short-term indebtedness to protect farmers from foreclosures and to aid farmers in regaining possession of property that had already been foreclosed, (3) provide federal assistance to farmers and their creditors in reaching agreements to scale down the level of rural indebtedness to meet the realities of the depression, (4) provide for the liquidation of Joint Stock Land Banks, and (5) provide new capital for the Federal Land Bank system.

The same parliamentary procedures that had been used on the farm bill were again adopted in the House to insure the quick passage of the credit bill. Debate was limited to eight hours and no amendments to the credit bill were to be given consideration. With nearly 40 percent

Many Middle Western farmers refused to abandon their farms without a fight. This angry farmer took up arms to protest eviction. (State Historical Society of North Dakota)

of the nation's farmers having debts totaling nearly $12 billion, and with demands for relief flooding Washington, the Administration's credit bill gained immediate and widespread support. The only speech in the House against the bill was by Representative Gerald Boileau of Wisconsin, who wanted inflation, rather than improved credit, to lift farmers out of the depression. On April 13 the bill passed the House by an overwhelming vote of 387 to 12. Of the twelve who voted against the bill, only two, Boileau and Farmer-Laborite Ernest Lundeen, who favored the Frazier-Lemke bill, were from the Middle West. The House had moved quickly to pass both the

Agricultural Adjustment Act and the Emergency Farm Credit Act; the Senate proved to be a more serious obstacle to Roosevelt's plans for agricultural relief.

The Senate, which considered the farm bill and the credit bill simultaneously, refused to be stampeded by the agricultural emergency. An immediate effort was made to amend the House bill by giving the Secretary of Agriculture discretionary power to guarantee farmers cost-of-production. John Simpson had persuaded Senator Norris to introduce the amendment in the Senate Agriculture and Forestry Committee. When the amendment passed in the committee, Simpson tried to convince Roosevelt and Wallace to accept the cost-of-production proposal. Simpson, who believed that the Democratic platform (which had promised that the President would support programs that gave farmers prices in ''excess'' of production costs) had endorsed his cost-of-production campaign, wrote Roosevelt that he was only performing ''my full duty to you, to the farmers of the Nation and to the party they have trusted. I am a firm believer in the principles of the Democrat Party as expounded by Jefferson and Jackson and I hope to always be found doing my part to preserve its integrity.'' With strong support from the Farmers' Holiday Association, Simpson also labored to convince Wallace that he should support cost-of-production. In spite of increasing pressures, Wallace continued to actively oppose the amendment.

On April 11 Wallace wrote to Roosevelt, ''I am convinced there is no satisfactory yard stick by which to measure cost of production. Each farm differs. Moreover, there are many different methods of keeping books. Also there are important differences of opinion as to what constitutes cost of production.'' Wallace still hoped to find a compromise that would satisfy the left and the right, but as debate on the bill continued he quickly discovered that his middle-of-the-road position failed to pacify conservatives, who believed that the bill went too far, and infuriated radicals, who believed that the bill did not do enough for the American farmer. His good friend in Iowa, Donald Murphy, wrote to the harassed Secretary, ''By this time you know how a bone feels when there is a dog at each end.''

In spite of Wallace's activity the Administration's fight against the cost-of-production amendment failed; on April 13, by a vote of forty-seven to forty-one, the Senate voted to include the Norris-Simpson amendment in the farm bill. Only four Senators from the

Middle West voted against the cost-of-production amendment. Simpson was satisfied with the amendment and continued to urge Wallace to discontinue his opposition to the Senate version of the bill. To win the Secretary's support, Simpson told Wallace on April 5 that he would be satisfied if the cost-of-production provisions of the farm bill were applied to only one crop.

In addition to the continuing fight over cost-of-production, tremendous pressure was placed on Roosevelt to add an inflationary amendment to the farm bill. By mid-April over forty inflationary bills, with considerable support from business as well as agriculture, had been introduced in Congress. Some inflationists wanted to do something about silver, others favored devaluation, while still others proposed printing paper money. On April 27, with strong support from Senators from the Middle West, Senator Elmer Thomas of Oklahoma introduced an omnibus measure authorizing the President to remonetize silver, devalue the gold content of the dollar, or issue greenbacks. Neither Wallace, who only a year earlier had believed that inflation would do as much for the farmer as anything else, nor Roosevelt, who was relatively flexible on the money question, was completely opposed to the Thomas amendment, but they did not want inflationary pressures to dominate the New Deal's approach to the farm crisis. When it became apparent that the Thomas amendment had united the forces pushing for inflation in the Senate, Roosevelt agreed not to oppose the amendment if its powers were discretionary rather than mandatory. Thomas agreed to revise his amendment, and on April 28, by a vote of sixty-four to twenty-one, the inflationary amendment was added to the Senate version of the farm bill. Only four Senators from the Middle West, Robert J. Bulkley of Ohio, Roscoe Patterson of Missouri, Thomas D. Schall of Minnesota, and Arthur Vandenberg of Michigan, voted against the Thomas amendment.

There was still considerable opposition to the farm bill from conservative Republicans, Eastern Democrats, and processors of agricultural products; but on April 28 the bill, with added inflationary and cost-of-production amendments, passed the Senate by a vote of sixty-four to twenty. Again, only four Senators from the Middle West voted against the bill. The bill was then sent to a conference committee of the Senate and House where disagreements on the farm bill were to be worked out.

While Congress debated, violence again erupted in the corn belt. At Le Mars, Iowa, Judge Charles C. Bradley was carried from his courtroom by an angry mob of farmers when he refused to promise that he would not preside over any more mortgage foreclosures. Taken to a crossroad outside of town, Bradley was beaten, a greasy hubcap was placed on his head, and his pants were filled with dirt. When he still refused to meet the farmers' demands, a noose was placed around his neck and he was jerked off his feet until he nearly lost consciousness.

As violence spread in Iowa, Governor Clyde Herring declared martial law in a half-dozen Iowa counties and called up the National Guard to maintain order. F. H. Shoemaker, a Minnesota Farmer-Laborite, telegraphed Herring that the troops should be sent home, adding, "If government were not used to fasten a system of slavery on farmers there would be no battle." H. R. Gross and Tom White, of the Iowa Farmers' Union and the Farmers' Holiday Association, insisted that none of their members had taken part in the violence, but Milo Reno attacked Herring for siding with farmers during election years and then turning troops against them after he was safely in office. Reno said he regretted the violence, but had decided it was inevitable. The Farmers' Holiday leader concluded, "I fear the Republic is facing, by far, the gravest danger in its history . . . and while I hope we will be able to solve our problem without a disastrous and barbarous revolution, yet I fear that those who have determined to destroy the Republic will force the people to drastic measures."

While the state of Iowa was more radical than the remainder of the Middle West, isolated incidents in a number of other states convinced many agrarian radicals that the farm community had developed a sense of solidarity and had arrived at a revolutionary stage of class consciousness. On May 3 the Farmers' Holiday Association opened a national convention in Des Moines, Iowa, with nearly fifteen hundred delegates in attendance. Outside observers contended that the Farmers' Holiday Association represented no more than 1 percent of the farm population, but the Association claimed to represent from 1 to 1.5 million members. The convention demanded a cost-of-production guarantee for farmers and an interest rate of only 1.5 percent on farm mortgages to be refinanced by the government. If their demands were not met, the convention voted to go on strike May 12. Reno wrote to Wallace on May 8 suggesting, as he would many times, that the Secretary of Agriculture resign or change his position on cost-of-production. On the same day, the Minnesota Farmers' Holiday Association, meeting in convention in Montevideo, Minnesota, also voted to call a farm strike. The Minnesota farmers resolved not to pay any debts until the dollar reached an honest measure of value, demanded that the government take over all banking facilities, and asked Roosevelt to declare a moratorium on foreclosure sales. The Minnesota members of the Farmers' Holiday also passed a resolution demanding that Roosevelt remove Wallace from office.

With Norris and Simpson continuing the legislative battle for cost-of-production, and with Reno threatening a national farm strike, many feared that even a small spark would ignite an agrarian explosion in the Middle West. Few agrarian radicals were prepared to test their popularity against Roosevelt, but many were convinced that the President was being mislead by his advisors, or as Reno preferred to call them, a bunch of "educated nuts." While the Farmers' Holiday Association was undoubtedly exaggerating its strength, no one could be certain in the spring of 1933 of the Association's real or potential following in the farm belt.

In mid-April Dr. Arthur E. Holt, a professor of Rural Sociology at the University of Chicago, completed a two-thousand-mile trip through Wisconsin, Minnesota, Iowa, and the Dakotas. After talking with representative farmers and their leaders, Holt warned that unless Roosevelt's program was immediately effective a general farm strike in the Middle West was a certainty. C. B. Miller, president of the Iowa Farmers' Union, wrote Roosevelt in early May that unless the cost-of-production feature were left in the farm bill, "we believe revolution is inevitable." When Edward O'Neal, of the American Farm Bureau Federation, released the results of a personal survey which indicated that farmers in the Middle West would not support a strike, Reno responded that the report was "exactly what might be expected from pussy-footed politicians and servants of special interest." Roosevelt took the threat of revolution seriously but refused to be intimidated by Reno or Simpson.

Roosevelt and Wallace continued to pressure the Congress to reject the cost-of-production amendment. On May 9, still following the President's lead, the House voted 283 to 108 to reject the Norris-Simpson amendment to the farm bill. Of the 108 who refused to yield to Administration pressure, 55, including 34 Democrats, 17 Republicans, and 4 Farmer-Laborites were representatives from the Middle West. The next day the Senate voted, 48 to 33, to withdraw the Norris amendment from the farm bill. Of those who voted against withdrawing the amendment, fourteen Senators, including influential Lynn Frazier of North Dakota, Robert La Follette, Jr., of Wisconsin, Peter Norbeck of North Dakota, George Norris of Nebraska, and Henrik Shipstead of Minnesota, were from the twelve Midwestern states. With the battle for cost-of-production at least temporarily over, Roosevelt's farm measures were quickly passed in the House and Senate.

On May 12 Roosevelt signed the farm bill, which included the Agricultural Adjustment Act, the Emergency Farm Mortgage Act, and the Thomas Amendment, into law. The bill was a major victory for the Roosevelt Administration and the farm community. It included most of the proposed solutions to the farm problem that had been debated since the 1920s. The Administration had refused to accept the cost-of-production proposal but had yielded on demands for inflation and made a vast commitment to improve the farm credit system.

It is difficult to see how the cost-of-production proposal would have improved the plight of the poor in the United States. Roosevelt's farm program may have been based, as Reno charged, upon a philosophy of economic idiocy, but the primary goal of the cost-of-production plan, like the domestic allotment proposal, was to raise farm prices, not to eliminate the contradiction of want in the midst of plenty. Agrarian radicals were most effective when they pointed to starvation and the enduring problem of underconsumption in the economy, but cost-of-production was not an alternative to the market price system of capitalism. They also failed to develop a better plan to deal with the problem of overproduction. They weren't blind to the issue, but they were simplistic and unimaginative. They talked about giving food to the poor but did not deal with the reality that the agricultural surpluses could not all be given away. Such an approach also did nothing to attack the root causes of poverty or to confront the serious maldistribution of wealth in the country. Attacks on the rich may have been appealing to the poor, but inflation, lower freight rates, cheap credit, and higher farm prices promised nothing to the urban poor. By failing to unite the urban and rural poor in a truly radical coalition, agrarian radicals ultimately surrendered the field of planning and reform to the managers and "technocrats" they so despised.

Many farm spokesmen refused to accept the defeat of the cost-of-production amendment. Burton K. Wheeler of Montana contended that the Farmers' Union represented more dirt farmers than any other organization and asserted, "We ought to have the courage to stand up and express our own views and not take the dictation of some professor down there in the Department of Agriculture." Senator Lynn Frazier direly predicted that the defeat of the cost-of-production amendment would result in a gigantic farm strike. Milo Reno interpreted the defeat of the cost-of-production amendment as a determined effort to ignore the farmers' right to a decent standard of living. At the same time, however, Reno was searching for an excuse to call off the Farmers' Holiday Association's threatened strike. When Minnesota Farmer-Labor Governor Floyd B. Olson asked the Holiday to give Roosevelt's farm program a chance, Reno used the governor's request as a pretext for calling off the strike. Although Reno's reasons for wanting to postpone the farm strike are obscure, he was disturbed by the amount of violence that had erupted in the farm belt and perhaps was fearful that sentiment for a farm strike was also declining. Whatever his reasons, Reno had not forsaken his demand that farmers were entitled to cost-of-production. The farm strike was not over: before the end of the year Reno would be heard from again.

The farm bill had given the Secretary of Agriculture such diverse discretionary powers that no one could predict with certainty which relief proposal the Administration would follow. One Middle Western farmer succinctly captured the confusion when he said, "I have read over the new farm bill exactly nine times and I can't tell for the life of me whether it'll turn out to be a reaper or a Ferris Wheel." Farmers in the Middle West, remembering the debacle of Hoover's Farm Board, waited to see if the new farm legislation would bring the improvement that Roosevelt and his advisors had promised.

The government provided a variety of work relief programs for farmers throughout the 1930s. Others relied upon government loans, cash grants, and surplus food distribution programs for their survival. (Nebraska State Historical Society)

Chapter 3

The New Deal in the Countryside, 1933

The passage of the Agricultural Adjustment Act did not end debate about the future direction of the New Deal's farm policy. The controversy was simply transferred from the halls of Congress to the Department of Agriculture. The Secretary of Agriculture, Henry Wallace, and his Assistant Secretary, Rexford Tugwell, insisted that production controls were necessary to restore prosperity to the nation's farmers. They disliked the prospect of cutting back production while millions of American families were living on near-starvation diets, but, with American foreign markets gone, they reluctantly concluded that the domestic allotment plan was the only feasible method of increasing agricultural prices. President Roosevelt, however, hopelessly divided the Administration when he appointed George N. Peek, who had previously championed McNary-Haugenism, to head the newly created Agricultural Adjustment Administration (AAA).

Peek, along with many of his contemporaries, conceded that production controls, which he labeled a "Hoover policy," might be necessary in rare circumstances as an emergency measure. As a permanent policy, however, Peek concluded that limiting production to domestic demands would "reflect unfavorably upon commerce, transportation, and particularly our social system." Peek continued to insist, as he had throughout the 1920s, that the answer to the farm crisis was to be found in marketing agreements with processors to raise prices, a high protective tariff, and, most important, the expansion of American foreign markets. Wallace and Tugwell believed that marketing agreements had a promising future but doubted that the United States could increase farm exports without relying upon export dumping, a program which they vigorously opposed.

Peek was also soon at war with his legal staff over the question of whether the AAA should be used as an instrument of social reform to improve the plight of the rural poor. In spite of Peek's objections, Jerome Frank, who had strong support from Tugwell, was named General Counsel for the AAA. Frank, whom Felix Frankfurter described as having "a fiendish appetite and capacity for work," recruited a number of talented and ambitious liberal reformers, including Thurman Arnold, Adlai Stevenson, Alger Hiss, Lee Pressman, John Abt, and Nathan Witt, to work with him in the AAA. They immediately split with Peek on the purpose of the AAA's marketing agreements. Peek believed that the agreements had but one purpose — to raise prices. Frank and his followers, however, believed that processors paid farmers too little for their products. It was the government's responsibility, they argued, to redress the imbalance by channeling a greater percentage of the profits from agriculture into the pockets of farmers.

Peek was certain that Roosevelt intended that he should be in control of the farm program. He later asserted that it was well understood that the broad powers granted to the Secretary of Agriculture by the farm bill "were going to be exercised only nominally by Messrs. Wallace and Tugwell and that the actual administration would be in the hands of someone with mature experience, guided by an unofficial advisory board of men in whom the public would have confidence." Wallace was equally certain that Roosevelt wanted the Secretary of Agriculture to be in charge of the Administration's agricultural program. In spite of appeals by Peek and Wallace that Roosevelt decide who was the boss, the President refused to intervene. When asked about the dispute at a press conference on May 19, Roosevelt replied, "I never heard of any, it is news to me."

The internecine warfare smoldered under the surface as the New Deal farm program took shape in the summer and fall of 1933. Peek's plans to open the channels of international trade met with almost complete failure. In October Peek was able to arrange for the export of twenty-eight million bushels of wheat, grown in the Pacific Northwest, to the Far East. A subsidy was paid on the exports at a cost of nearly $6 million. The sale of the wheat was possible only because the Reconstruction Finance Corporation (RFC) had loaned China $10 million with the stipulation that over half of the money had to be used to purchase American wheat. When Peek asked the Secretary of

Agriculture for an additional $500,000 to subsidize the export of butter to Europe, Tugwell, who was acting for Wallace, refused the request. Wallace still hoped to convert Peek to his ideas, but he became increasingly worried that it might be impossible to handle the stubborn and persistent Peek. In October Wallace wrote to Dante Pierce, ''George has many admirable qualities but his personal prejudices are terrible when once aroused. On the other hand he is a splendid bull dog when he has once taken a stand on the right things. He is a little slow in understanding the facts of the situation.''

By November Peek was become increasingly exasperated with the ''professors'' and ''intellectuals'' in the Department of Agriculture. A number of marketing agreements had been put into effect but had not had the dramatic results Peek had anticipated. The only marketing agreements of great importance in the Middle West, those for the milk marketing industry, were already breaking down. In late November Peek, who blamed the legal counsel for creating a ''bottleneck'' in the AAA's marketing agreements program, tried to remove Frank from office. Tugwell was especially incensed by Peek's efforts to ''purge'' the AAA. He wrote in his diary, ''I had a kind of fondness for the bluff, shrewd and persistent Peek. But these qualities were unfortunately consistent with an essential stupidity which finally became intolerable.'' Peek had forced Roosevelt's hand; the President had to choose between Peek or Wallace and the liberals in the AAA.

Roosevelt's habit of putting people with opposing philosophies together, in the hope that they would find a middle ground, obviously did not work in agriculture. Still, he was reluctant to dismiss Peek, who remained popular with farmers, particularly in the Middle West. Fearful that Peek would lead a crusade against the New Deal's farm program, Roosevelt decided to keep him in the New Deal family by ''promoting'' him to the position of president of the Export-Import Bank and Special Advisor on Foreign Trade. Roosevelt thus resolved the dilemma of conflicting philosophies about the direction of the AAA by assigning Peek, who still wanted a high tariff for agriculture, to work with his Secretary of State, Cordell Hull, who was striving to lower tariff barriers throughout the world. They too would soon be locked in mortal combat.

Peek continued to speak out against Wallace, Tugwell, and Frank. Their exchanges were increasingly bitter and personal. They did not

Many federal relief projects were designed to be labor intensive to provide the unemployed and the destitute with work. Here workers complete a farm pond and spillway near Bellwood, Nebraska, in 1935. (Nebraska State Historical Society)

like each other, and they disagreed almost completely on the role the government should play in agriculture. Equally important was the question of whether managers and planners in Washington, old-line progressives who wanted government powers carefully defined and narrowly circumscribed, or farmers at the grassroots level would decide United States farm policy.

Peek had long championed federal regulation of the economy to guarantee farmers "equality of opportunity," but he balked at policies that would limit the farmers' freedom of choice. Increasingly Peek saw the New Deal as a conspiracy to lead farmers down the path of socialism. The plan, as Peek saw it unfolding, "was first to undermine the independence of the farmer through putting him on the dole and, when that had been done, to regulate agriculture exactly as the Departmental bureaucrats saw fit." Peek argued that he and his supporters, like Charles Brand, who served as coadministrator of the AAA for a brief period in 1933, were kept in the AAA as "window dressing" and that neither Congress nor farmers would have accepted the Agricultural Adjustment Act if they had known that Wallace and Tugwell would be in control. Milo Reno joined in the attacks on government bureaucrats, charging that Peek's removal was still more

evidence that ''nothing short of a complete dictatorship for agriculture will satisfy the administration.''

The battle to determine the future of American agriculture now
moved to the nation's farms and ranches. Every organized group,
from the left to the right, insisted that it represented the true voice of
agriculture. Few farmers, however, and especially few poor farmers,
belonged to any of the major farm organizations in the 1930s. The
AAA program had been carefully packaged as a farmers' bill, but it
still had to be sold to the farm community.

Much of the early effort of the AAA was directed toward explaining the New Deal farm program to farmers in order to win their
support for controlled production. Wallace was careful to emphasize,
however, that while acreage reduction was the immediate objective
of the Administration, the AAA would also pursue the long-range
objectives of opening new international markets and improving
land-use practices in the United States. He also tried to persuade
farmers that they must accept an expanded role by the government in
the farm economy. He warned, ''The ungoverned rush of rugged
individualism perhaps had an economic justification in the day when
we had all the West to surge upon and conquer; but this country has
filled up now, and grown up. There are no more Indians to fight. No
more land worth taking may be had for the grabbing. We must
experience a change of mind and heart.''

While officials in Washington blitzed the countryside with a
massive ''educational'' campaign, the AAA developed its first
acreage reduction program for wheat producers. As one of the major
wheat-producing areas of the world, the Middle West was obviously
crucial to the Administration's plans. Wheat growers and processors
were among the most united and best-organized farmers in the nation.
They were also among the most desperate. With a wheat carryover of
939 million bushels on July 1, 1933, and with wheat exports, which
had totaled 313 million bushels in 1920-1921, having fallen to only
32 million bushels in 1932, the domestic allotment plan met with
little immediate resistance in the Middle West.

Since the wheat crop had already been planted, AAA officials first
considered plowing under wheat acreage to prevent an even larger
surplus in 1934. Ultimately they decided that a plow-up campaign
was not necessary because dry weather in a number of Plains states
promised to naturally limit wheat production in 1933. For 1934 and

1935, however, it was decided that benefit payments, financed by processing taxes, would be paid to farmers who agreed to cut back their wheat acreage by an amount determined by the government. The 1934 contract called for a benefit payment of twenty-nine cents a bushel to be paid to farmers who reduced their wheat acreage by 15 percent of their average acreage from 1930 to 1932. Nationally 77 percent of the nation's 1.2 million wheat producers signed contracts. Of the major wheat-producing states in the Middle West, 91 percent of the wheat farmers in Kansas and 93 percent in North and South Dakota signed up for the wheat reduction program. Although the contracts did not affect the 1933 crop, it was decided that the first allotment payments, equal to about two-thirds of the total payment for reducing production in 1934, would be paid in the fall and winter of 1933.

Secretary Wallace and AAA officials like M. L. Wilson, who headed the wheat section of the AAA, insisted that farmers themselves should have an active role in administering as much of the AAA program as possible. They also realized, however, as much as they might want a bold new experiment in democratic planning, that they could not create a new administrative hierarchy overnight. But it was impossible, politically and economically, to delay the implementation of the New Deal's farm program. The result was a makeshift administrative structure that would tie the New Deal farm program to the past as well as the future.

Prior to the New Deal, if farmers had any direct contact with the federal government, it was usually through the land-grant colleges that had been created after the Civil War to disseminate scientific information to the nation's farmers. In 1914 the Smith-Lever Act was passed to provide funds, on a dollar-matching basis, to support the colleges' increasingly important ''extension'' activities. The money provided by the states to match the federal contribution could be from legislative appropriation or from donations from private groups. The new law also provided that plans for extension work had to be approved by the Secretary of Agriculture before they could be put into effect.

The results of the dollar-matching programs were uneven; the poorer states usually lagged behind and the lines of authority were frequently confused. To coordinate their programs, and to consolidate their new power, many of the land-grant colleges began to appoint state extension directors whose nominations were, at least in theory, subject to approval by the Department of Agriculture. In 1923, as the number of extension directors proliferated, a federal director of extension services, who was directly responsible to the Secretary of Agriculture, was appointed to coordinate the activities of the various state extension services. Since the federal director shared his responsibilities with the state extension directors, his power and authority were usually limited to a supervisory role.

Field representatives of the state extension services were usually referred to as "county agents." The local and state groups which organized to raise dollar-matching funds to support the county-agent system were frequently referred to as "farm bureaus." When these quasi-private, quasi-state, quasi-federal associations began to federate into state farm bureaus, they emphasized that their primary function was education, not politics. By 1919-1920, however, when the state farm bureaus federated to form the American Farm Bureau Federation (AFBF), the potential for exercising political power became irresistible. Since county agents received part of their funding from the federal government, they were frequently viewed as "public" officials representing the Department of Agriculture. At the same time, however, county agents received funding from "private" farm bureaus and often felt compelled to support the ideological positions of the national federation.

The AAA turned to the agricultural extension service for assistance in creating community, county, and state production control associations to administer the New Deal's farm program. To decentralize the administration of its programs, the AAA created community committees to make acreage allotments and to make certain that farmers complied with their contracts. The extension services, through the county agents, were usually directly involved in supervising the elections and in choosing the committee members. In some cases, although the members were supposed to be elected, the first community associations were simply appointed by the county agents. The community committees, which were almost always dominated by members of the AFBF, then sent representatives to form county associations. State boards were then chosen, usually by state AAA directors, who were chosen either by the state extension directors or by officials of the AAA in Washington.

The close relationship between the extension service and the AFBF

The Civilian Conservation Corps gave work to young men in the country as well as in the city. In the Middle West, many CCC camps worked with farmers to restore the land and develop sound conservation techniques. (Nebraska State Historical Society)

distressed many supporters of the Farmers' Union and the Farmers' Holiday Association. The extension service had never been popular with the nation's poorest farmers, in part because the funds used to finance the county-agent system frequently came from the business community, but more importantly because the new technology, which the extension service passed on to the nation's farmers, only further undermined their precarious economic existence. Poor farmers, who could not afford the cost of modernizing their farms, blamed the county agents for creating the problem of overproduction in the first place. Many marginal farmers were also convinced that the county-agent system, which appeared to be on the verge of collapse as local funds began to dry up during the depression, would be the primary beneficiary of the AAA program. Milo Reno spoke for many enemies of the extension service and the AFBF when he wrote, ''In my opinion, Henry Wallace intends to perpetuate the county agent system that has been the curse of agriculture, taking away the power of the people to eliminate them, and making them a part of the federal political machine. . . .''

Wallace was not unmindful that the decision to rely upon the extension service would create a storm of controversy in the Middle West. Even before he became Secretary of Agriculture, Wallace had heard recurrent complaints that the county agent was about as popular ''as a skunk would be under your Ford.'' Still Wallace concluded that while many county agents were ''rather dumb,'' most of them had ''their hearts decidedly in the right place.'' Wallace admitted that the domestic allotment plan would probably serve to rescue the financially troubled county-agent system, but he also believed that the extension service would eventually benefit every class in American agriculture. More important, it also enabled the AAA to begin activities without delay.

The AAA wheat program was well underway before the Administration took action to improve the plight of corn and hog producers. Like farmers in the wheat belt, corn and hog producers faced declining prices, a lessened domestic demand, the loss of their foreign markets, and economic bankruptcy. For a number of years before 1930 the yearly slaughter of hogs had brought farmers an annual average income of over $1 billion; by 1932 their income had fallen to less than $450 million. Hogs, which had sold for $7.22 per hundredweight in the years 1910-1914, sold for only $3.00 per hundred pounds in the winter of 1932. Corn prices had fallen from sixty-four cents a bushel in the years 1910 to 1914 to only thirty-three cents a bushel in 1932. While corn and hog producers had exported nearly two billion pounds of pork and lard in the early postwar years, by 1932 their exports had fallen to only one-third that amount. Since 90 percent of the hogs and 75 percent of the corn produced in the United States came from the twelve Middle Western states, the corn-hog situation was critical to farmers in the Middle West.

The AAA had a much more difficult time developing a viable production control program for corn and hogs than for wheat. Since most of the corn produced in the Middle West was used to feed hogs, the corn-hog situation was infinitely more complex. Wallace, who worried about the political consequences of delay, later wrote, "Conferences by the score began and ended in my office, and as often as not many of us were ready to give up the problem as too tough." Corn and hog producers, unlike wheat producers, also lacked organizations to speak for them in negotiations with the AAA. By July 1933, with encouragement from Wallace, producers' committees were established in Iowa, Kansas, Nebraska, South Dakota, Minnesota, Ohio, Illinois, Wisconsin, Indiana, and Missouri. On July 18 the state committees met in Des Moines, Iowa, and selected a National Corn-Hog Committee of Twenty-Five to speak for farmers in the Middle West.

After consulting with the Committee of Twenty-Five and with representatives of the processing industry, AAA officials concluded that they could not wait until 1934 to reduce the supply of pigs and sows on the nation's farms. The AAA consequently agreed to purchase a maximum of one million sows, weighing not less than 275 pounds and due to farrow in the fall of 1933, and to purchase a maximum of four million pigs and lightweight hogs weighing between 25 and 100 pounds. The primary goal of the program was simple: to immediately improve the farmers' economic position by decreasing the number of pigs and sows that would flow to an already saturated market. The AAA also planned, however, to turn over the millions of pounds of pork and lard that it acquired to other relief agencies for distribution to the needy.

Unfortunately for Roosevelt and Wallace, the program was poorly understood from beginning the end. The AAA did an excellent job of selling its program to corn-hog producers; it failed, however, in the rush of events, to satisfactorily explain to the general public that food salvaged during the operation would be given away to the poor. Wallace, who had argued before he became Secretary of Agriculture that the public simply would not tolerate the destruction of food while millions of Americans were hungry, was quickly charged with "murdering" six million little pigs. Father Charles Coughlin, the Detroit radio priest, spoke for millions of perplexed Americans when he wrote to Roosevelt's Secretary, Marvin McIntyre, "I am truly chagrined at these foolish proposals aimed at starving us into prosperity." John Simpson, warning that it was impossible for the nation to destroy its way back to prosperity, wrote to Roosevelt, "Your 'Brain Trust' has justified my statement that they are a bunch of impractical theorists. They attempt to hitch up the agricultural economic horse and cart; and just as I prophesied, they got the cart before the horse."

Many farmers in the Middle West were also upset by the prospect of "wasting the country back to prosperity," but were eager to participate in any program promising agricultural relief. Newspaper correspondent Roland M. Jones summarized the sentiments of many farmers when he observed, "It is pretty hard to work up resentment against one who comes asking instead of commanding and offers to pay for what he wants in the bargain."

From August 23 to October 7 the government purchased 6,410,866 pigs and sows, at a cost of $30,643,101.95. Many of the pigs and sows were destroyed, some because they were unfit for human consumption, others because they were too small to be handled by the processors' dehairing machinery, and still others because of the lack of storage facilities. But nearly 100 million pounds of salt pork and 8 million pounds of lard, representing about 40 percent of the total purchases, were salvaged. Another 22 million pounds of the government's purchase was converted into inedible grease or into dried fertilizer. From an economic perspective the pig purchase program made sense; from a public relations perspective, it was a disaster. Throughout the remainder of the decade the AAA would have to contend with the image that it wantonly destroyed food that should have been used to feed the nation's poor.

While the emergency pig purchase program was underway, a plan was also announced to prevent future overproduction. Farmers who reduced their corn acreage by up to 20 percent of the base period, and

Farmers in the 1930s were still a powerful political force which few politicians failed to recognize. Farmers staged countlesss protests in the early 1930s to demand government aid. Here farmers march on the state capitol building in Lincoln, Nebraska. (Nebraska State Historical Society)

the number of litters farrowed and the number of hogs marketed from those litters by as much as 25 percent, were to be given benefit payments for cooperating with the government. December 2, 1931, to November 30, 1932, was chosen as the base period. As with the wheat program, the production control program was to be administered by local production committees composed of farmers and their representatives.

The third basic agricultural industry in the Middle West, the dairy industry, also faced the problem of increasing production and declining prices. The index number for dairy prices fell from 140 percent of the prewar average in 1929 to only 59 percent in March and April of 1933. Still, none of the representatives of the dairy industry favored production controls. Rather, they wanted the government to purchase surplus dairy products, particularly butter and cheese, to stabilize prices. For the long run they proposed that the government use marketing agreements to fix retail prices and to increase the prices farmers received from their distributors.

Wallace insisted that some form of production control was needed. Obviously exasperated, the Secretary observed, ''There would seem to be something about the dairy business which leads a man to bury his head in the flank of a cow and lose track of time and space.'' When dairy spokesmen refused to bend, the AAA capitulated and agreed to buy surplus dairy products as a ''temporary measure.'' In the summer and early fall, marketing agreements were also negotiated in Chicago, St. Paul-Minneapolis, Evansville, Des Moines, and St. Louis. By the end of the year the agreements, which proved to be extraordinarily difficult to enforce, began to break down. In its annual report the AAA made it clear that production controls were needed, and warned, ''No short cuts or makeshifts, no resort to expedience or temporizing would absolve the dairy industry from ultimate action on production adjustment in some form.''

While the AAA continued to try to persuade farmers to sign production control contracts, the Farm Credit Administration (FCA) began to work in the Middle West. The Emergency Farm Mortgage Act and the supplemental Farm Credit Act of June 1933 promised farmers relief by scaling down both the interest and the principal of their mortgage indebtedness. For the more than four hundred thousand farmers who already held Federal Land Bank loans, the Emergency Farm Mortgage Act reduced the interest rate on their loans from an average of 5.5 percent to 4.5 percent. The new credit bill also provided for the issuance of $2 billion worth of 4 percent bonds to refinance mortgages held by creditors other than the Federal Land Banks. To be eligible for the loans, however, the amount of the farmer's mortgage could not exceed 50 percent of the normal value of the land and 20 percent of the value of the insured improvements. Since a large percentage of farm mortgage indebtedness was held by nongovernmental financiers for more than 50 percent of the normal value of the land, many farmers could benefit from the bill only if private mortgage-holders agreed to scale down the principal of the mortgages they held. If mortgage-holders refused to cooperate, farmers could, provided their indebtedness was not too high, still apply for a new mortgage loan from the Federal Land Banks to pay off their existing indebtedness.

For those farmers who had already lost their land or were about to be foreclosed, the bill also promised relief. The RFC made $200 million available to the Farm Loan Commissioner to make loans to farmers who had been foreclosed after July 1931. For those who were barely hanging onto their farmsteads, loans were available to pay off outstanding debts and to carry on farming operations. The loans, which could be made up to the amount of $5,000 at 5 percent interest, were not to exceed 75 percent of the appraised normal value of the land.

On May 10, 1933, Henry Morgenthau, the governor of the FCA, announced that he was calling a conference of the leading life insurance companies, which held an estimated $3 billion worth of farm mortgages, to try to persuade them to exchange their farm mortgages, at a depreciated value, for Federal Land Bank bonds. On May 15 a disappointed Morgenthau, after meeting with thirty-three representatives of leading insurance companies, announced that the industry was opposed to any general reduction in mortgage values. Expecting an upturn in the economy, most of the private holders of mortgaged property believed that if they waited long enough they would be able to collect the full amount of the principal that was due them.

Rebuffed by the private sector, the FCA launched an attack on another front. Many banks in rural areas were still closed, creating tight credit conditions where credit was needed most. To remedy the situation the FCA began an experimental program in Wisconsin to refinance farm mortgages which were locked up in closed state banks. The FCA made $35 million available to purchase first mortgages from closed or restricted banks in Wisconsin. Since the refinancing was limited to one-half of the normal appraised value of the land, many rural debts were automatically scaled down. The program, which was eventually applied to other states in the Middle West, thus scaled down the farmers' burden of indebtedness, and the new funds also enabled many rural banks to reopen on a solvent basis.

Farmers who had already lost their farms, or who needed operating capital to prevent foreclosure proceedings, were less fortunate. The process of applying for and receiving money from the government took time — time the farmers did not have. On July 22 Morgenthau announced that farmers had already applied for $100 million of the $200 million fund to save farmers from foreclosure, but the money had only begun to flow into the hands of farmers. After John Bosch, vice president of the Farmers' Holiday Association, telegraphed Roosevelt that foreclosures by banks and insurance companies were proceeding at a rapid rate, Morgenthau publicly appealed to mortgage-holders to have patience and to delay foreclosure proceedings until the federal refinancing program had time to become effective. Again Morgenthau was disappointed. The "army of the homeless," as John Simpson called the nation's dispossessed farmers, continued to grow at an alarming rate. Roosevelt and Morgenthau were determined to save the capitalist order. Capital, however, seemed determined to do everything possible to fan the fires of discontent in the farm belt.

The emergency pig purchase program, the creation of AAA Committees, the sign-up campaigns, and the emergency credit activities of the government offered hope, but by the summer of 1933 the era of good feeling that followed Roosevelt's inauguration was nearing an end. The farm community was still badly divided and economic suffering was still widespread. Amy O. Rogers, of Missouri Valley, Iowa, wrote to Louis Howe, Roosevelt's minister without portfolio, "We are just 'boiling inside.' I use both sides of this paper to save postage. My house letters go on government postal cards. All I can buy with prices as they are and we are told to 'buy in August!" With What? Tell me!"

By the middle of August, as the AAA worked around the clock to hurry its first allotment checks to the farm belt, Wallace again expressed concern about the political climate in the Middle West. He worried that farmers were still not committed to production controls and that most still believed that inflation was the only thing that was really needed to raise farm prices. Reno again launched a vicious attack against Wallace, charging that it was evident "that no relief can be hoped for until Secretary Wallace's position is held by some man, who is really the Secretary of Agriculture instead of the tool of the Wall Street bunch." Wallace warned the President that, with the farm organizations being pressed "by the violent holiday group and other left-wing organizations," the agricultural situation would come to a head sometime in September. The New Deal had arrived in the Middle West, but revolution still did not seem out of the question.

An Agrarian Revolution?

By the late summer of 1933 the New Deal ceased to be a vague collection of ideas and began to have a tangible meaning to farmers in the Middle West. In both agriculture and industry the New Deal had

Members of the Cass County, North Dakota, Farmer's Holiday Association gather to prevent a farm foreclosure in the 1930s. (State Historical Society of North Dakota)

launched a massive campaign to raise prices by limiting production. The National Recovery Administration (NRA), which had been created to help business and labor work together to raise prices and wages through production codes, enjoyed early success. During the fall industrial prices, including the cost of products which farmers had to buy, such as farm machinery, rose sharply. Farm prices, however, lagged behind and were still depressed. The AAA promised that production controls would also raise price levels for agricultural products, but farmers needed relief immediately, not at some unforeseeable date in the future. Talk of strikes, price-fixing, cost-of-production, and inflation again reverberated throughout the Middle West.

The rise in industrial prices was interpreted by rural militants as an effort to save big business and Eastern financial interests at the expense of the agricultural population. John Simpson, president of the Farmers' Union, tried to persuade farmers that the nation's big bankers had captured control of the NRA and that only the wealthy would benefit from the New Deal's farm program. C. N. Rogers, who also spoke for the Farmer's Union, warned Roosevelt that unless farm buying power was restored the Administration would have to cope with a revolution in the countryside. Even the more conservative farm organizations appeared on the verge of withdrawing their support from the Administration's farm program. The Iowa Farm Bureau Federation urged Roosevelt to fix prices for corn and hogs by governmental fiat and asked the President to abandon voluntarism by making production controls mandatory. Hans Pauli, one of the Wallace's close friends, warned that the patience of the masses was exhausted. He concluded on an alarmist note, ''It is my candid opinion, Henry, that the allotment plan will be buried long before it will be of any benefit to the farmers.''

Pressures to increase farm prices immediately by promoting inflation also attracted widespread support. Even Edward O'Neal, president of the American Farm Bureau Federation, joined in the chorus of requests that Roosevelt cheapen the value of the dollar. In a radio speech on September 9 O'Neal argued, ''The 3-horse chariot of the administration — the A.A.A., the N.R.A. and inflation must pull together for national recovery or the team will be destroyed. The N.R.A. has been pulling faster than the A.A.A. and inflation has been put to work scarcely at all. The A.A.A. must put on more speed and inflation must be made to press on the collar if we are to reach the goal of national prosperity.'' While O'Neal pressed for quick action on inflation, Father Coughlin, who had a tremendous following in the farm belt, also began to pressure Roosevelt to begin an inflationary program to bring immediate relief to farmers. Coughlin insisted that he was still with Roosevelt ''one million percent,'' but he was in fact quickly losing his enthusiasm for the President and the New Deal.

Most of the criticism of the New Deal farm program was directed not at Roosevelt, who remained personally popular, but at his Secretary of Agriculture, Henry A. Wallace. Milo Reno mounted the offensive against Wallace in the corn belt. The president of the Farmers' Holiday Association called Wallace a ''jackass'' and an ''ignoramus'' and contended that he had never known a member of the Wallace family who had not been mentally unbalanced. On September 20 the Iowa Farmers' Union, which was meeting in convention in Des Moines, joined in the attack and demanded that Wallace resign as Secretary of Agriculture. Wallace was not surprised by the assaults on his character, but he wrote, somewhat gloomily, ''I must confess that deliberate misquotation and misunderstanding pain me. . . .'' Wallace still had his defenders, but their voices were drowned out by the rising tide of discontent.

As the crisis deepened Wallace warned Roosevelt, ''There is a great storm now brewing,'' and advised the President that he expected the situation to become ''exceedingly tense'' during the next several months. Wallace still vigorously defended production controls but reluctantly concluded that it might also be necessary to adopt price-fixing or inflation in the near future. Roosevelt, too, was forced to publicly support the Administration's farm program. At a press conference on September 13 the President pointed out that although farm prices were not nearly high enough, gross farm income in 1933 would be more than a billion dollars higher than in 1932. In a letter to the governor of Iowa, Clyde Herring, Roosevelt expressed the hope that it would be possible to coordinate the activities of the AAA and the NRA and pledged to support changes in the farm program if, by midwinter, the Administration's policies had been proven inadequate. Roosevelt cautioned the Iowa Governor, however, that ''price-fixing of a type which builds up stocks and maintains acreage for a product of which there is no effective demand either at home or abroad, is to be avoided because of certain failure later, no matter how expedient it may seem to be for the time being, politically.''

The desperate plight of many farmers in the Middle West was heightened by drought conditions and by a massive invasion of grasshoppers. Harry Hopkins, administrator of the Federal Emergency Relief Administration (FERA), indicated that in South Dakota alone from forty to fifty thousand people would be on relief the entire winter as a result of drought. The FERA, which had been created in May 1933 with powers to make grants to the states for direct relief, promised to provide food, clothing, work relief, and feed for livestock in the drought areas. Harold Ickes, the administrator of the Public Works Administration (PWA), which had already begun an extensive program of public works, indicated that road work in the drought areas would be extended to employ needy farm families. Henry Morgenthau, Jr., of the FCA, also promised to liberalize federal credit policies in the drought areas. In spite of the government's emergency programs, farmers in the Middle West faced the winter of 1933 in a mood of angry despair.

Farmers who continued to fight foreclosure were especially bitter. By September the FCA had begun to receive numerous complaints that it was too slow in implementing the provisions of the Emergency Farm Mortgage Act and that when it did act it appraised the value of farm property at unreasonably low levels. By the middle of September the FCA had a backlog of 185,000 applications and was receiving 17,000 additional applications for loans to refinance farm mortgages each day. The FCA obviously did not have enough trained personnel to handle the volume of demands for relief, and again the Administration was charged with siding with bankers and insurance companies against farmers. Milo Reno summarized the confusion, despair, and dissatisfaction with the FCA when he wrote to Eleanor Roosevelt, ''Men and women, who have built the farm homes in this middle west — patriotically and happily contributing their services to society — are in their old age being dispossessed and thrown out on the world, helpless and hopeless, through no fault of their own.''

The FCA's program to purchase farm mortgages from closed rural banks also met with unexpected obstacles. Many rural banks, which had made unsound loans on inflated values before the depression, could not qualify for the federal program and remained indefinitely closed. Like other New Deal farm programs, the FCA promised relief in the future but could not meet the farmers' demands for immediate relief.

Many Middle Western farmers had been willing to give Roosevelt a chance but were now convinced that the New Deal farm program was a tragic mistake. Representative Edward Eicher of Iowa informed President Roosevelt that he had been flooded with letters, telegrams, and telephone calls, "all sounding the doleful note that not a single one of our party's undertakings designed to help the farmer has proven so far to be anything other than a disappointment to the people of Iowa." F. E. Murphy, publisher of the Minneapolis *Tribune*, complained that Roosevelt had gotten the cart before the horse and that as soon as the rural economy revived there would be business recovery. Representative D. C. Dobbins of Illinois wrote the President that the farmers of his state were convinced that the Farm Mortgage Act would not bring relief, that the AAA was being handled in a too "professional manner," that the NRA worked against their interests, and that, although money was flowing into banks, farmers were unable to get any of it. John Simpson, still hoping to push Roosevelt to the left, confidently predicted, "President Roosevelt must change his policies and do it very soon or he is a lost soul politically."

The Department of Agriculture's argument that overproduction was the primary cause of the agricultural depression seemed all the more ludicrous to discontented farmers. Deeply religious, many agreed with Father Coughlin, who advised Roosevelt that he should get rid of "that asinine philosophy propagated by Henry Wallace which dares to contest the Lord's Prayer of 'Give us this day our daily Bread.' " Milo Reno charged that it was "criminal" and "un-American" to destroy food. Reno wrote to S. P. Free, "I am certain that their program cannot succeed, because it is wrong and any program that is not right, its ultimate end is failure."

While many New Dealers, including Henry Wallace, were sincerely religious, the Roosevelt Administration appealed for support on scientific rather than moral grounds. Whether fundamentalist religious groups would support political movements on the right or the left was still uncertain in the early days of the New Deal. Past experience, however, particularly during the First World War, convinced many reformers that moral crusades resulted in hysteria and reaction as often as constructive change. It was also clear, however,

Iowa farmers march on the state capitol during the winter of 1933 demanding that the legislature act to provide agricultural relief. (State Historical Society of Iowa)

that unless the New Deal responded to charges that the farm program was un-Christian, it was doomed to failure. New Dealers were modern, not stupid. They prided themselves on being pragmatic, but they were as distressed as their critics by the enigma of starvation and overabundant agricultural production.

Roosevelt tried to change the image that the Department of Agriculture was dominated by callous, unthinking, insensitive men when he announced, at a press conference on September 21, that the AAA would buy surplus food and clothing for the needy unemployed. The food and clothing were to be turned over to the FERA, which in turn gave the surplus commodities to individual state relief administrations for distribution. For greater efficiency in purchasing and distributing the products, a new federal agency, the Federal Surplus Relief Corporation (FSRC), was created on October 4, 1933.

Giving surplus food to poor people was not new. Surplus products had been given to the poor when Hoover was President, and the FERA gave away food as a part of its general relief program. The FSRC would, however, always be poorly funded and was always limited in its scope. Although the government had more food than it knew what to do with, the Roosevelt Administration was always fearful that if people were given food they would become permanent wards of the government. Many people, including most of the major farm organizations, still blamed the poor for their condition and worried that public relief would undermine the moral fiber of the nation. Thousands of people depended upon FSRC food for survival throughout the 1930s, but many found direct food relief to be the most degrading form of public assistance. Professional social workers, preferring that the poor be given work or direct cash grants, looked to food handouts as a last resort and failed to push for an expanded food giveaway program. The FSRC's primary mandate, especially after it was placed under the direction of the Department of Agriculture in 1935, was to benefit farmers by reducing surpluses, not to guarantee that the poor and hungry had adequate diets.

In spite of the creation of the FSRC, Roosevelt's popularity in the Middle West continued to decline. Even in the wheat belt, which had been so receptive to the domestic allotment plan, there were rumblings of discontent. Wheat farmers who had agreed to reduce production in 1934 became angry and discouraged when their wheat allotment checks failed to arrive on time. Roy Buckingham observed in Dodge City, Kansas, that the farmer who signed up for the program in good faith was now in the "mood to believe he has been sold another gold brick." From North Dakota Governor William Langer, who had previously championed inflation and cost-of-production proposals for wheat farmers, urged Roosevelt to fix a minimum price for wheat. He insisted, "If it was the proper thing to do during the World War when times were prosperous it is needed thousands of times more when we are in the midst of this crisis."

With farm products due to move into the hands of processors within sixty to ninety days, nearly all of the major farm organizations tried to convince Roosevelt that he could not ride out the crisis by simply reiterating the promise that the Administration's program would bring relief in the future. On September 25, representatives of the farm and livestock industries, including Edward O'Neal, E. A. Eckert of the National Grange, C. E. Huff of the Farmers' National Grain Corporation, Charles A. Ewing of the National Livestock Marketing Association, D. N. Geyer of the National Cooperative Milk Producers Federation, and farm editors Clarence Poe and Dan Wallace, called on Roosevelt and asked that he take steps to raise the price of their commodities by supporting inflationary measures that would reestablish an "honest dollar." Less than a week later the Corn and Hog Committee of Twenty-Five, which had been created to help the Administration sell the idea of "voluntary" production controls, placed a proposal before Secretary Wallace which called for the government to fix prices for both corn and hogs.

The more militant Farmers' Holiday Association and the Farmers' Union also met with Roosevelt to discuss their demands. On October 7 Roosevelt met with John Bosch, Harry C. Parmenter, C. F. Lyttle, and R. S. Noon, all representing the Farmers' Holiday Association, and E. E. Kennedy, secretary of the Farmers' Union. The group pressured Roosevelt to declare a moratorium on foreclosures until Morgenthau's program became effective, urged the President to consider a code of fair practices for agriculture, and insisted that the price of all farm commodities be fixed at cost-of-production. They further threatened that if their demands were not met, they were authorized by their parent organizations to call a farm strike.

Roosevelt faced pressure from still another front when Governor Langer of North Dakota announced on October 16 that he had declared an embargo in the state on the sale of spring wheat at less than cost-of-production prices. In spite of a sharp decline, wheat prices were higher than those for meat, poultry, and dairy products; but Langer insisted that "the time has come to bring the problem of the farmers' welfare to a head if recovery is to become a reality." Other Middle Western governors, however, refused to join in the embargo. Without their support, the effort to stop wheat from flowing to market was doomed to failure. Langer's gesture failed to raise prices, but the governor did succeed in further dramatizing the farmers' discontent in the northern Plains states.

Publicly Roosevelt continued to support the New Deal farm program. On the same day that Langer declared the embargo, however, Roosevelt telephoned Henry Morgenthau, expressing alarm over the price of wheat. He told Morgenthau, "I can't take it any longer. . . . Can't you buy 25,000,000 bushels for Harry Hopkins and see if you can't put the price up?" Peek, Morgenthau, and Hopkins soon worked out a plan which called for the Farmers' National Grain Corporation, a private agency that had worked closely with Hoover's Federal Farm Board, to purchase thirty million bushels of wheat. The wheat in turn was given to Hopkins for relief distribution. Wheat purchases, supervised by Morgenthau, began immediately. On the first day of the wheat purchase program Morgenthau jubilantly observed, "We accomplished what the President wanted, and I felt that this was one of the big moments of my life. I let the President know around noon how things were going, and I told him that the firm of Hopkins and Morgenthau were in the wheat business." The purchasing activities momentarily stabilized the price of wheat and bought the administration time. Still, as Hoover had learned earlier, emergency surplus removal operations by the government were not a solution to the farm problem.

Just as Wallace had predicted, other leaders increasingly looked to inflation to end the farm crisis. Charles Bryan, the governor of Nebraska, sympathized with Langer's embargo in North Dakota but worried about whether an embargo was constitutional. Milo Reno, who hoped that Bryan would support the Farmers' Holiday Association, encouraged Bryan to abandon all restraint and assured him that the constitution had "been shot all to Hell long ago." Like his famous brother, William Jennings Bryan, the Nebraska governor was a determined enemy of big business. He was convinced that the NRA was cutting the farmer's throat "from ear to ear" and that big business was profiteering at the expense of the masses. Bryan encouraged Roosevelt to support inflation, and he was joined by Nebraska's best-known Senator, George Norris, who exercised tremendous influence and was widely respected in Washington. Norris telegraphed Roosevelt on October 18 asking him to issue new money, instead of new securities, to retire a $1.5 billion issue of Liberty bonds. Norris defended the farm organizations' demands for inflation, writing, "These associations have not exaggerated the situation. The farmer is rapidly losing confidence because he sees the price of everything he buys going up, while the price of everything he sells is daily going the other direction."

In mid-October George H. Adams, editor of the Minneapolis *Star*, warned James Farley, Roosevelt's Postmaster General, that the Northwest was ready for an explosion. Attention still focused on the Farmers' Holiday, but reports also filtered in that the Communist Party had begun an aggressive campaign to win support in the Middle West. The Communist Party gained a few converts, but its success was quite limited. Farmers wanted relief and were willing to make unusual alliances to accomplish their goals, but few were willing to abandon the concept of private property or to believe that capitalism was beyond redemption. Still, the collapse of the economy encouraged farmers to engage in violence against the "system" and brought a new willingness to listen to left wing attacks against privilege, wealth, and the nation's business elite. If the government, and the traditional farm organizations, failed to meet the farmers' demands, many feared that the Communists might try to seize power as the Bolsheviks had in Russia in 1917. In the uncertain and emotionally charged atmosphere that pervaded the countryside during the fall of 1933, Communist activities were considered a serious revolutionary threat.

The Communist presence in the Middle West also made it difficult for Reno and the Farmers' Holiday rank and file to moderate their demands. After attending a meeting of the board at a Farmers' Holiday Association gathering in Des Moines in September, Leon Vanderlyn, chairman of the Associated Liberals, wrote to Louis Howe, Roosevelt's secretary, that the Communists were anxious to see Reno discredited. Vanderlyn indicated that if Reno compromised, the Communists were ready to take over the farm protest movement. Reno, who was convinced that the Communists were being financed by "powerful eastern groups," deplored the movement but was being swept along by events that appeared to be out of control. He, too, took seriously the possibility that farmers in their desperation would be driven into the arms of the far left.

Reno's hatred of the New Deal farm program grew with each passing month. On October 21 he, along with twenty other directors of the Farmers' Holiday Association from Minnesota, North Dakota, South Dakota, Wisconsin, and Iowa, issued a call for a national farm strike to begin immediately. Contending that the government was in the hands of bankers, racketeers, and monied interests, the directors of the Farmers' Holiday called for the refinancing of the farm debts and asked the government for measures that would inflate the value of the dollar. The primary objective of the strike, according to Walter Groth, secretary of the Minnesota Association, was to force the government to approve a code for agriculture, comparable to the NRA codes for industry, which would guarantee farmers at least cost-of-production.

Old issues quickly surfaced. Reno charged that Wallace's program was a bribe from the same old dealers using the same stacked deck, and contended that the strike would determine whether the farmer would retain his independence or would "become a peasant, the menial slave of the usurers and the industrialists." John Simpson, who endorsed the strike, linked together the old order with the new planners and managers who now controlled the Department of Agriculture. He informed Roosevelt, "I'm afraid your 'Brain Trust' is not keeping you informed of what the crooks are doing to you. I suspect a number of your 'Brain Trust' have been in the employ of the crooks in the past and may not have severed all of their connections." Both Reno and Simpson promised to work with Roosevelt to guarantee farmers cost-of-production, but they warned, "We are going to strike until agriculture is given its rightful place in the economic and social sun of this nation."

When the strike began, Reno claimed to have two million members. It was soon apparent, however, that the farm community was badly divided about whether the strike was necessary. Edward O'Neal, along with other AFBF spokesmen, condemned the strike. C. A. Dyer, legislative agent for the Ohio Farm Bureau and the Ohio Grange, suggested that Reno had overestimated his strength. Dyer concluded, ''The Ohio farmers are dissatisfied, but they are not radical and won't follow that windy group in the west.'' Still, a number of Middle Western governors initially responded with favor to the strike. Governor William Langer of North Dakota and Wisconsin Governor A. G. Schmedeman, announced that they would support the strike. The governors of Iowa and Minnesota, Clyde Herring and Floyd B. Olson, expressed general agreement with the objectives of the striking farmers.

Secretary Wallace disliked the bitter spirit of the strike, but he understood the need for more immediate relief in the farm belt. George Peek went even further when he said, ''All these people are trying to do is to save their homes and save their property. I too would fight to hold my home. We have been warning the East for twelve years that things like this would happen unless the incomes of farmers were increased.'' President Roosevelt was considerably more alarmed than either Peek or Wallace. Fearing that an agrarian revolution might be underway, the President took immediate action to undermine Reno and his followers.

The New Deal Responds

During the fall of 1933, as the farm strike began to unfold, President Roosevelt took several steps to head off a possible revolution. During a ''fireside chat'' on October 22 Roosevelt assured the nation's farmers that he was still their friend. The President promised to do whatever was necessary to raise prices and announced that the Administration's immediate goal was fifty-cent corn and ninety-cent wheat by early 1934. Upset by charges that the Administration was controlled by international bankers, Roosevelt publicly appealed to the financial community to postpone mortgage foreclosures until the refinancing program could benefit farmers. Most important, in a move opposed by the conservative banking community, Roosevelt also announced a new gold purchase policy to respond to the farmers' demands for inflation.

Following the advice of Professor George Warren and Secretary of the Treasury Henry Morgenthau, Roosevelt was persuaded that if the government purchased gold at increasing dollar prices, the dollar would be devalued and prices would rise. Also, it was hoped that the devalued dollar would in turn stimulate American exports. Ultimately the gold purchase program would fail to promote inflation or raise farm prices, but the move won Roosevelt renewed support in the Middle West. Father Coughlin wired Roosevelt his congratulations, George Peek jubilantly interpreted the gold policy as a direct concession to farmers' demands, and John Simpson praised the President's bold and decisive action. While some members of the Administration opposed the inflationary policy because it avoided the question of production controls, Roosevelt was convinced that the purchase of gold, along with the earlier announcement that the government would purchase surplus wheat, had prevented a revolution. At a meeting on October 29 he told his financial advisors, "Gentlemen, if we continued a week or so longer without my having made this move on gold, we would have had an agrarian revolution in this country."

Earlier in 1933 the Commodity Credit Corporation (CCC) was created as a corollary to the AAA production control program. Under the program, when the AAA failed to limit production as much as was desired, the CCC made loans to farmers on products they stored on their farms. If the price of the commodity rose, farmers could reclaim the product, repay their loans, and then sell the stored products on the open market at prices above the loan value. If the price did not rise, the product became the property of the government to dispose of as it saw fit. The Administration now moved to extend the activities of the CCC to the corn belt. AAA officials were anxious to pump more money into the corn belt to undermine the farm strike. They feared, however, that farmers would resort to unlimited production if they were guaranteed minimum prices by the loan program, and that the government would be saddled with a tremendous loss.

Wallace resolved the dilemma when he announced on October 25 that the government would extend the CCC program to include corn to those farmers who agreed to participate in the government's acreage reduction program. According to the plan outlined by the AAA, farmers would be given a loan of forty-five cents per bushel of corn at 4 percent interest in those states that had Farm Warehouse Acts, if farmers agreed to reduce their production of corn by 20 percent and would reduce by 25 percent the number of litters farrowed for market. The program was a modified version of price-fixing, but, unlike the gold purchase program, it was not well received in the Middle West.

William Hirth prophesied that the new commodity loans would

make an already bad situation worse. John Simpson complained to the President that the farmers needed to be able to pay off their existing debts, not to borrow more money. Simpson, who also opposed the repeal of prohibition, again warned Roosevelt not to follow the unsound advice of his ''Brain Trust.'' Summing up the New Deal recovery program as one of drink, borrow, and destroy, Simpson chastised the President: ''There is not a ten year old farm boy in the United States so ignorant as to believe anyone can drink himself into prosperity, borrow himself into prosperity, or become prosperous by destroying property.'' Simpson and other opponents of production controls still hoped to exploit the farm crisis to win Roosevelt over to their point of view. The Administration, however, continued to refine its response to the farm crisis.

The Farm Credit Administration also announced a new initiative to speed up its mortgage refinancing program. On October 20 Morgenthau announced that he had asked Middle Western governors to set up state, county, and local conciliation boards to aid farmers in scaling down their level of indebtedness. The debt adjustment committees arranged for debtors and creditors to meet together to discuss adjusting the farmers' indebtedness to meet the realities of the depression. Although participation in the program was voluntary, some creditors, especially in rural areas, were anxious to collect on their debts, even at reduced values. Because of the new program, many farmers whose level of indebtedness had excluded them from the FCA program before were now able to benefit from the refinancing provisions of the Emergency Farm Mortgage Act.

While Roosevelt orchestrated the powers of the government to bring immediate relief to farmers in the Middle West, the much-feared strike got underway. Reno had called the strike on October 21, but by the end of the first week only a small number of farmers in parts of Iowa, Minnesota, and Wisconsin were actively participating. Militant rhetoric again filled the air as Reno and Simpson exhorted their followers to join them in battle. In Shenandoah, Iowa, a mass meeting of the Farmers' Holiday spanked an effigy of Wallace, but it was soon apparent, even in Reno's home state, that enough farmers were opposed to the strike to make any withholding action ineffective. Roosevelt's quick response to the perceived threat of violence and revolution played a major role in stalling the strike movement. After the fireside chat on October 22, the Farmers' Holiday Associations in western Nebraska and Colorado voted down strike proposals. Other agrarian spokesmen, including George Norris and Charles Bryan of Nebraska, indicated that they were now pleased with the Administration's response to the farm crisis and were hopeful that the agricultural situation would soon improve.

Still, Roosevelt, along with state government officials, worried that the Middle West remained a powder key just waiting to explode. In late October Governor Clyde Herring of Iowa invited the governors of North Dakota, South Dakota, Nebraska, Kansas, Missouri, Indiana, Wisconsin, Minnesota, and Illinois to attend a conference on the farm crisis. Three hundred persons, including representatives from the Farmers' Holiday Association, the Farmers' Union, the American Farm Bureau Federation, the Wisconsin Cooperative Milk Pool, and Governors Tom Berry of South Dakota, A. G. Schmedeman of Wisconsin, Willliam Langer of North Dakota, and Floyd Olson of Minnesota, attended the conference.

Farm Bureau spokesmen urged the conference to support the New Deal's agricultural relief program, but the governors were obviously fearful of Reno and the strike movement. Before the conference adjourned, Governors Berry, Langer, Herring, Olson, and Schmedeman endorsed a report asking for an NRA code for agriculture, more inflationary measures, an expansion of the government's mortgage refinancing program, and a program of minimum price supports for agricultural products. They also announced their intention to go to Washington to plead their case with Roosevelt personally.

On November 1 the five governors, in what Wallace later called "one of the most interesting thunderstorms [he] ever watched," arrived in Washington for a three-day series of conferences. At their first meeting with Roosevelt they presented him with a code to fix agricultural prices. Obviously referring to the AFBF, the governors blamed "fictitious farm organizations, sponsored and supported by ambitious bureaucrats affiliated with the federal government or chambers of commerce," for creating "a condition of blind competition in production to the utter neglect of a controlled sales system." To respond to the crisis, the governors advocated compulsory marketing controls, with each farmer given a definite quota to sell on the market. Only three years earlier the Middle West had been a conservative Republican stronghold. Its most influential political spokesmen now called for a degree of regimentation that would all but abandon agriculture's traditional "free" market system.

President Roosevelt received ominous reports during the conference that striking farmers were blocking roads and blowing up milk and cheese factories. Roosevelt hesitated, but he ultimately decided to reject the price-fixing proposal. Edward O'Neal, with obvious relief, and Secretary Wallace, who opposed the plan as unenforceable and too costly, commended him on his stand. The Farmers' Union and the Farmers' Holiday, however, charged that Roosevelt had broken his preelection pledge to support cost-of-production. For the first time since he began his campaign for the Democratic nomination in the spring of 1932, Roosevelt had been coerced into stating publicly that cost-of-production plans were unworkable and would not be supported by his Administration.

The governors' reactions to Roosevelt's decision varied. Berry indicated that he was still with the President; Schmedeman worried about the future; and Olson, who emerged from the conference with a national reputation, still insisted that something had to be done to promote recovery. Langer was angry and disgusted. The North Dakota governor bitterly concluded, "Everybody else, the banker, the insurance man and the railroad man, was here ahead of the farmer and got their money. There is nothing left for the farmer. The farmer is the forgotten man."

Reno hoped that Roosevelt's stance would win additional support for the farm strike. He had placed little faith in the governors and was still convinced that the strike would force Roosevelt to accept his demands. Early in November Reno wrote to George W. Christian, "I believe that there is enough opposition to the present system that has been a complete failure as far as human happiness is concerned, to overthrow the powers of darkness that now control and restore the government of the United States to the people who have builded it and sustain it." For a few brief moments, it appeared that Reno might be right. On November 7 the Minnesota Division of the Farmers' Union passed a resolution demanding that Roosevelt give them a Secretary of Agriculture who would not "advocate policies to destroy the blessing of life so long as our Brothers are starving and in need of clothing." Representative Shoemaker, a Farmer-Laborite from Minnesota, telegraphed Roosevelt that farmers were organizing military units under the command of former servicemen and that fighting between pickets and antistrikers was increasing with alarming frequency.

Within a week, however, it was obvious that Reno, the governors, and Roosevelt, had seriously overestimated support for the strike. There continued to be scattered violence in Iowa, Minnesota, and Wisconsin, but the strike was virtually stillborn. Farmers who had previously remained silent now openly voiced their approval of Roosevelt and his policies. Others formed ''counter-revolutionary'' groups to oppose the strike. On November 8 at Le Mars, Iowa, a group of farmers met to form a ''law and order league'' to keep the roads open. They also adopted a resolution supporting Roosevelt and denouncing Milo Reno. Three days later over five hundred farmers met outside Sioux City to perfect plans for an organization to oppose the strikers. At Watertown, North Dakota, twelve hundred farmers met to form a ''protective association'' to oppose the strike movement. Father Coughlin, temporarily convinced that Roosevelt was not controlled by international bankers, also rallied support for Roosevelt in the farm belt.

Before the revolt had time to renew its vitality, Roosevelt decided to send General Hugh Johnson, head of the NRA, and Secretary Wallace on a pacification tour of the Middle West. Johnson spoke with powerful eloquence in defense of the activities of the NRA. He told farmers they were sadly misled if they thought the NRA was the major source of their troubles, and pleaded with them to be patient. Wallace, in an address before ten thousand people in Des Moines, told farmers they could join the Midwestern governors, they could ''raise hell'' with the Holiday Association, or they could cooperate with the government. Seeking support for a middle ground, he argued that the hell-raisers and the reactionaries were working toward the same end, since neither had given the Administration a chance. He warned the impatient crowd, ''If the Administration can be pushed into doing something foolish, the reactionaries can come into their own again.'' In Muncie, Indiana, Wallace again explained that until farmers could regain their lost foreign markets, it was necessary to limit production to meet domestic demands. Wallace told his audience that he did not want to return to ''the vomit of capitalism,'' but again appealed for moderation. He assured his listeners, ''I believe there is a middle course by which we can shake off the leadership of discredited capitalists without committing ourselves to the follies of the hellraisers.'' Reno again charged that Wallace was a liar and a tool of Wall Street, but the Secretary had persuaded many farmers that the Administration was concerned about their welfare and was earnestly trying to solve the farm problem.

Harry Hopkins, the head of the Federal Emergency Relief Administration, added to the compassionate image of the Administration by increasing direct relief payments to farmers. The FERA focused its attention on the urban unemployed, but by the fall of 1933 there were already over one million people in rural areas of the Middle West who were dependent upon public relief for survival. Lorena Hickok, one of Hopkins' most perceptive field observers, reported terrible suffering in rural areas, particularly in the drought-stricken Dakotas. She wrote of South Dakota, ''It is the 'Siberia' of the United States. A more hopeless place I never saw.'' She reported that many farmers, who had not harvested any crops for four years because of drought and grasshoppers, were in desperate need of food and clothing, and had only soup made of Russian thistles to feed their families in the coming winter. She accurately concluded, ''I have an idea that the chief trouble with the people of South Dakota, the thing that is behind whatever unrest there is in the state, is sheer terror.'' In many areas the FERA's first relief efforts were disorganized and poorly funded. In spite of relief programs, many farm people and animals were near starvation. Hickok reported that in Morton County, North Dakota, nearly one-third of the one thousand families in the county, most of which were farm families, were living on relief allowances of only six dollars a month.

In early November Hopkins announced the creation of the Civil Works Administration (CWA) to provide work relief for four million people throughout the United States during the winter months. The creation of the CWA was particularly significant in the Middle West. By the third week in November 253,385 people in the Middle West were employed on CWA projects building public roads, public buildings, and parks, or working on erosion control projects. By the middle of December the number had risen to 998,366. The government also announced that it would continue purchasing surplus pork and dairy products from the market to distribute to the needy. From November 1933 until May 1934 pork products equivalent to about 60,000 head of hogs were eventually purchased at a cost of nearly $15 million. By January 15, 1934, the government had spent

$11,250,000 for the purchase of butter and cheese.

As the CWA went into operation, money from the government's corn loan program and the AAA's wheat reduction program began to flow into the Middle West. The administrative confusion in the FCA was also ironed out; farmers now found it much easier to borrow money to refinance farm mortgages and to seed the next year's crops. While many Middle Western farmers still questioned the value of the domestic allotment plan, the influx of immediate cash benefits broke the back of whatever resistance remained to Roosevelt's farm program. A random poll by the Des Moines *Register* in November showed that only 17 percent of the farmers interviewed were in favor of the NRA and that only 37 percent favored the corn loan program. The poll also indicated, however, that 77 percent were against the farm strike.

It was obvious that the farm revolt was dead. On November 18 Langer lifted the wheat embargo in North Dakota. The next day Reno called off the farm strike. By the end of the year the mood of the Middle West had changed from desperation to buoyant optimism. Roosevelt, too, seemed confident that the Administration's farm program was at last on its way. On December 4 the President wrote to Edward O'Neal, ''The Middle West has had a long hard time of it. It has submitted with an admirable patience to necessary delays in relief. Better times are more than due for the great central region of our country, and I am confident now that better times are coming there.''

Farm prices improved somewhat by the end of the year. Farm cash income in 1933 totaled $5,591,000,000 compared with only $4,377,000,000 in 1932. Much of the improvement was due to widespread drought. In 1933 an estimated sixty-seven million acres of land had been planted to wheat, but, as a result of the drought, only forty-seven million were harvested. Wheat production fell from an annual average in the 1920s of 860 million bushels to only 529 million bushels. Wheat exports, which had totaled 313 million bushels in 1920-1921, had fallen to only 32 million bushels. Still, the cash income from wheat production rose from only $196,000,000 in 1932 to $275,360,000 in 1933. The average price of wheat, which had been only thirty-eight cents in 1932-1933, averaged seventy-four cents a bushel for the 1933-1934 crop. In addition the rental and benefit payments paid to wheat farmers in the Middle West for promising to reduce their acreage during the next two years totaled $13,502,061.12.

The income from corn and hog production, which had fallen to only $597 million in 1932, or nearly $850 million less than in 1929, also improved slightly in 1933. The price of hogs had risen from $3.44 per hundredweight in 1932 to $3.94 per hundredweight in 1933. Corn, which sold for an average of 28.1 cents a bushel in 1932, sold for an average of 38 cents per bushel in 1933. Another important source of income for corn and hog farmers in 1933 was the corn loans from the CCC. Ultimately $120,493,034 was loaned to 197,000 farmers on 267,761,708 bushels of corn under the 1933 program. Almost all of the money was loaned to farmers in the rebellious cornbelt of the upper Middle West. In addition the emergency hog purchase program in August and September had put millions of dollars in the pockets of Middle Western farmers.

Income from dairy products in 1933 was $938 million, compared with an average from 1925 to 1929 of nearly $1.7 billion. The AAA's marketing agreements had had mixed results, but dairy prices had risen to 82 percent of the prewar average by the end of the year.

Many farmers were still heavily in debt, but the farm credit situation was much improved. In December 1933 the FCA loaned $98 million, more than the total of federal loans in the previous two and one-half years. Eighty-five percent of the Federal Land Bank Loans and 90 percent of the Land Commissioner Loans granted by the FCA in 1933 were used to refinance farm mortgages. Nationally 43,408 Federal Land Bank and Land Commissioner loans had been made by November 30, 1933, totaling $125,046,567.21. Of this total, 20,926 loans were made in the Middle West, totaling $65,686,135.14. In addition to the Federal Land Bank and Land Commissioner loans, 80,727 Emergency Crop and Feed Loans, totaling $9,174,816, were made to farmers in the Middle West.

The New Deal farm program was still an unproven experiment. The government had pumped millions of dollars into the economy of the Middle West, but Roosevelt had still not won the support of the agrarian left for the voluntary domestic allotment plan. In December Simpson wrote to Reno complaining of the Administration's "underhanded" tactics. He warned Reno, "Most of the mean ones under Hoover are still in, and to these have been added a lot of new mean ones." Reno, although he had called for a dramatic increase in governmental power, accused the New Deal of trying to destroy the country. He wrote to J. W. Tabor, ''The old problem of the divine right of a few to rule and exploit the many is again the issue and if the liberties of our people are maintained and perpetuated for your children, it will be by the determination, patriotism and sacrifice of ourselves.''

Differences between the agrarian left and the New Deal were exaggerated by the atmosphere of crisis, the farm strike, and Reno's rhetorical attacks against the capitalist system. The left's charge that Roosevelt was saving the capitalist system by giving aid to banks, insurance companies, and big business was true. Indeed, one of the primary legacies of the New Deal was the salvation of corporate capitalism. Roosevelt knew that the system needed significant reform, but he still believed that the capitalist order was fundamentally sound and just. He did not launch an attack against wealth and privilege, but at the same time it was impossible to argue that he had turned his back on farmers. One might question whether farmers were getting their fair share. There remained fundamental differences about how best to restore prosperity to the farm economy. There were honest doubts about whether the New Deal's farm program would work. By the fall of 1933, however, none could deny that the federal government had made a massive commitment to save agriculture as well as business and industry. The New Deal promised balance, not revolutionary change.

The agrarian left may not have liked Washington's ''bureaucrats,'' but they expected the government to solve the farm crisis. They called for more, not less government. At first glance they were highly critical of the capitalist system. Their demand that the government fix prices would have resulted in a more structured and regimented economy than liberals thought necessary. The willingness to risk violence through direct confrontation also separated agrarian radicals from New Dealers in Washington who urged patience and compromise.

There was more agreement, however, than either side realized in 1933. Radicals continued to exploit the issue of scarcity economics, but New Deal agencies were already giving surplus food, in limited quantities, to the poor. The agrarian left might hope for a redistribution of wealth in the United States, but their immediate demands for inflation, cost-of-production, and mortgage refinancing promised

only to protect the farmers' old order, not to bring about a revolutionary reconstructuring of society. Their goal was not to destroy the
system but to protect the farmers' right to own property and make a
profit. They sympathized with the poor, but offered them little. They
imagined a simpler age when the economy had been dominated by
farmers instead of by government and large corporations. They
realized that the ''wrong crowd'' had somehow gained control of
their lives, but they only vaguely understood the forces of modernization that had already begun to bring truly revolutionary change
to American agriculture. Simpson and Reno attacked the capitalist
system politically but failed to transcend the system morally or
intellectually to develop an alternative vision of the future. When
Roosevelt promised to save the system, and began to act on that
promise, he not only threatened the survival of the agrarian left, he
also exposed its intellectual bankruptcy.

Drought and hard times brought farmers together in cooperative efforts to help one another. They were also united in their demand that the government provide a comprehensive drought relief program. (State Historical Society of North Dakota)

Chapter 4

The Drought of 1934

In the opening months of 1934 rural militants renewed their attack against the New Deal's farm program. In early March, as Reno urged farmers to organize to demand "what rightfully belongs to them," the governors of Wisconsin, Iowa, and Minnesota again held a conference in Des Moines to discuss farm relief. The conference, which was attended by large delegations from the Farmers' Union and the Farmers' Holiday Association, still insisted that the Agricultural Adjustment Administration was a proven failure and again urged the Administration to adopt compulsory production controls to guarantee farmers at least cost-of-production. North Dakota Congressmen William Lemke, who emerged as one of the farmers' most articulate spokesmen, insisted that the New Deal had only managed to take "the drowning farmer out of seventeen feet of water . . . into an eleven-foot depth," where he was still drowning. Like Reno, he urged farmers to join a "militant block," to protect their homes, and to "tell the Agricultural Department to go to Hell."

Lemke also argued the farmers' cause in Congress. Throughout 1933 radical farm groups had agitated for the passage of a controversial bill to refinance farm indebtedness. The bill took several forms, but usually provided that the government would purchase all farm mortgage indebtedness and that farmers, who would be charged an annual interest rate of 1.5 percent, would have to pay the government only 1.5 percent of their indebtedness each year. To finance the program, the government would issue paper money, which, by promoting inflation, would scale down the amount of rural indebtedness to more "reasonable" depression levels. Roosevelt adamantly opposed the "wild" legislation and warned that it would destroy his recovery program. The bill died in committee.

Another bill sponsored by Lemke and by Senator Lynn Frazier,

A farmer looks to the heavens for rain. (Rothstein photo, State Historical Society of North Dakota)

also from North Dakota, did pass the Congress in June 1934. The Frazier-Lemke Farm Bankruptcy Act provided a virtual moratorium on mortgage foreclosures. The bill stipulated that if a farmer failed to reach an agreement with his creditor on debt adjustment, the farmer could declare bankruptcy, have the land appraised, and repurchase the mortgaged farm over a period of six years. If the farmer did not wish to begin purchasing the land immediately, he could remain on the land for a period of five years by paying nominal rent and interest to the mortgage-holder. On May 27, 1935, however, the Supreme Court declared the law unconstitutional.

While agrarian radicals attacked the New Deal for not doing enough for agriculture, representatives from the business community charged that Roosevelt had already gone too far. Thomas Y. Wickham, a leading spokesman for the United States Chamber of Commerce and the Chicago Board of Trade, criticized Roosevelt for

High temperatures and strong winds parched the earth and left farmers without enough water to survive. (State Historical Society of North Dakota)

turning to the "Brain Trust," instead of to merchants, who supposedly would have found farmers new markets without resorting to production controls. Food processors, especially those who were now bound by marketing agreements, complained of high taxes and unwarranted government interference in the market economy. Conservative farm spokesmen, who had been generally quiet during Roosevelt's first year in office, also expressed alarm over the general tone and direction of the New Deal. Sam R. McKelvie, a former member of Hoover's Farm Board and publisher of the *Nebraska Farmer,* hoped that anti-New Deal Congressmen would carry the elections in the fall. Until then, McKelvie pledged to keep his readers "informed of what is going on and to warn them of the great dangers that lie in the far reaching departures that have been made in our form of government.'' In spite of continuing opposition, in March 1934, the AAA broadened its production control program to include cattle, rye, flax, barley, grain sorghums, and peanuts as basic commodities.

The AAA's major challenge in 1934 would come not from political voices on the left or the right, but from the impersonal and unpredictable forces of nature. In 1934 one of the worst droughts in the nation's history visited the Plains, devastating virtually everything in its path. Drought was certainly not a new phenomenon to Middle Western farmers, but the drought of 1934, combined with the effects of the depression, presented a challenge without parallel to the nearly prostrate farm community. The ideological debate about who would control American farm policy continued, but in 1934 nature would dictate the New Deal's response to the continuing farm crisis.

The drought, which eventually affected three-fourths of the nation, began in the Northwest, spread to the Southeast and the deep South, and finally struck the Southwest. By February, reports from the upper Middle West indicated that many families were ''at the end of the road'' and that thousands of cattle were starving. By spring, as the drought spread, reports of water shortages, dust storms, feed shortages, insect infestations, and dying stock, particularly in Wisconsin, Minnesota, South Dakota, North Dakota, Kansas, and Nebraska, filtered into Washington.

On May 11 President Roosevelt announced that his Cabinet was already working on a program of drought relief. A. G. Black, Chief of the Corn-Hog section of the AAA, and George E. Farrell, Chief of the Wheat Section, were sent to the drought areas to gather information about the needs of drought-stricken farmers. Congressional delegations from the Middle West also began to place pressure on New Deal officials to increase public works and to expand federal relief programs in the most severely affected areas of their states.

The Roosevelt Administration's first response to the drought called for existing federal agencies to expand their activities in the drought areas. Harry Hopkins, administrator of the Federal Emergency Relief Administration (FERA), promised to continue providing funds to the various state emergency relief groups for work projects and to sponsor a program to purchase starving cattle. Chester Davis, who succeeded George Peek as administrator of the AAA in December 1933, indicated that crop adjustment contracts might be modified to allow farmers to plant summer forage crops. William I.

The drought cattle purchase program enabled the government to reduce the number of surplus cattle in the Middle West. Here cattle purchased by the government are buried in a mass grave. (Nebraska State Historical Society)

Myers, who became governor of the Farm Credit Administration (FCA) when Henry Morgenthau, Jr., was appointed Secretary of the Treasury in late 1933, promised to liberalize the terms of government loans from his agency. Officials in Washington also announced that railroads operating in the drought areas would be asked to reduce rates on cattle shipped from the drought area and on feed shipments into the drought-stricken regions.

To coordinate the government's relief activities the AAA established an emergency drought activities service, under the direction of E. W. Sheets. State drought relief directors were selected in a number of states, including North Dakota, South Dakota, and Minnesota. A system for rating counties as ''emergency'' or ''secondary,'' a designation which determined their eligibility for receiving special federal aid, was also established. Roosevelt had hoped that it would not be necessary to seek additional funding to meet the crisis, but by late May it was obvious that government agencies could not handle the additional burden. The FERA estimated that it would need at least $50 million to care for as many as two hundred thousand farm families during the next eight months. The FCA also indicated that additional funds, of at least $25 million, would be needed for

Many of the cattle purchased by the government during the 1934 drought were canned and distributed to the poor and hungry. Many of the cattle were unfit for human consumption and were immediately destroyed and buried on farms and ranches throughout the Middle West. (Nebraska State Historical Society)

emergency loans, to be handled by the FCA's Emergency Crop Loan Office.

Urgent requests for funds to buy feed for livestock and seed for quick-growing forage crops, and for additional work relief projects to increase and conserve water supplies, poured into Washington from the drought areas. In June rivers began to dry up and temperatures soared to record-breaking heights. By the end of summer the temperature had risen above 100 degrees more than fifty times in the Central Plains; on the hottest days the temperature neared 120 degrees. As the parched earth turned brown, the governors of North Dakota and Minnesota declared an embargo on cattle shipments into their states in an effort to preserve grazing land for hard-pressed farmers. The drama of the drought continued to unfold. Alfred C. Miller, of New Jersey, asked Roosevelt to declare a national day of prayer to call upon God to save farmers in the West; Senator La Follette warned, ''This drought is approaching the proportions of a national calamity.'' A. G. Crooker, of Fairmont, Nebraska, wrote to President Roosevelt, ''Most of the farmers around here and in the

Dakotas and other dry areas are utterly discouraged. There is nothing to look forward to but help from the government or some charitable millionaires.''

Farmers in many areas of the Middle West had to fight not only dry weather, but also a massive invasion of grasshoppers. The hoppers, which first appeared in South Dakota, Nebraska, and Iowa in 1931, plagued farmers throughout the decade. They moved in vast swarms as they devoured the few crops that remained. The 1934 government officials estimated that insects had caused $38 million worth of damage in the Great Plains region. When the crops were gone, the hoppers covered the sides of farmhouses, barns, and sheds, eating away the paint until the buildings had been stripped as bare as the land. The government provided poisoned bait and sprays to fight the hoppers, but it was not until 1941 that the plague was finally brought under control.

On June 2, as the drought worsened, President Roosevelt announced that the Administration would provide additional funds to provide work relief for farmers in the drought areas. Within three days 50,000 farmers were at work; another 150,000 were promised jobs by the end of the week. Roosevelt also met with his advisors and

The cattle purchase program resembled a wartime emergency as farmers rushed their livestock to market. Here ranchers in Verdigree, Nebraska, sell their starving stock to the government. (Nebraska State Historical Society)

with Congressional representatives from fourteen drought states, including Wisconsin, Minnesota, North Dakota, South Dakota, and Nebraska, to discuss the need for additional legislation. On June 9 the President sent a message to Congress asking for $525 million for a special drought relief program. The proposed bill provided $75 million for the purchase of beef and dairy cattle; $100 million to buy feed for livestock and to finance feed shipments into the drought areas; $25 million to buy wheat, corn, and forage seed to be planted the next year; $50 million for retiring land and for moving families out of the drought areas to better land; and $50 million for the expansion of the Civilian Conservation Corps. The bill passed Congress unopposed, and by the middle of June money for the drought relief appropriation began to flow into the Middle West. By the end of the year every state in the Middle West, with the exception of Ohio, would receive assistance from the government's drought relief program.

The drought brought a number of unexpected benefits to the New Deal's farm program. Issues that had divided the farm community and the Administration were put aside as the nation mobilized to save the land and the people of the Great Plains. The federal assistance program was hurried to farmers in the Middle West, as Rexford Tugwell later observed, almost with a sigh of relief. Tugwell recalled, "Here was something that could be done and done properly. It was important and worthy. Everyone moved to the attack with complete assurance of rightness."

Complaints from conservatives that the New Deal was trying to "regiment" the farmer were all but forgotten as farmers banded together to demand even more relief from the government. Opponents of the New Deal's planning efforts were forced to concede that the new bureaucracy in Washington could also bring benefits to Middle Western farmers. Arthur Capper, of Kansas, echoed the sentiments of many farmers when he observd, "Whether we like national planning for agriculture or not, the fact that we have it is going to help a lot in meeting this drought problem. The AAA forces, the extension services of the department of agriculture, the county agents . . . give the federal government the machinery with which to use the federal funds intelligently, effectively, quickly, where these are available and where these are needed."

Many farmers, schooled in the philosophy of "rugged indi-

Long-winged grasshoppers of the plains migrating during the heat of the day. (Nebraska State Historical Society)

vidualism'' and ''laissez-faire'' economics, felt confused, guilty, and ashamed when they first received relief payments from the government. Most had difficulty understanding and accepting the expanding role of the government in their daily lives. The drought, since it was a natural disaster beyond their control, made it easier to accept federal assistance, particularly if the assistance came in the form of work relief. It also made it easier to accept the New Deal and the need for a national program of agricultural reform.

The drought precipitated emergency programs in many areas with many federal agencies working in cooperation, but none proved to be more imperative or exhaustive than the efforts of the government to bring relief to livestock owners. From January 1, 1930, to January 1, 1934, the number of cattle in the United States increased by nearly thirteen million; the price of beef in 1930 had been $7.46 per hundredweight; by 1933 the price had fallen to only $3.63 per hundredweight. Cattle were not included as a basic commodity under the original Agricultural Adjustment Act. Ranchers, even more than farmers, were reluctant to accept production controls or regulations limiting the number of cattle they could graze on the Western range.

This cornfield has been badly damaged by grasshoppers. In other fields even the stalks were eaten. (Nebraska State Historical Society)

By late 1933 cattle producers, faced with declining demand and falling beef prices, began discussing the industry's problems with AAA officials in Washington. In April 1934, before the drought became serious, the Agricultural Adjustment Act was amended to include cattle as a basic commodity. Negotiations were still in progress about what should be done when the drought intervened to settle the issue.

By early summer water and feed supplies throughout much of the Middle West were severely diminished by the drought. In addition, many cattle had come through the winter of 1933 in poor condition. When the drought struck, many farmers and ranchers were forced to ship their cattle to market without regard to the prices they might receive; others, who could not afford to ship their livestock, watched helplessly as their cattle died on the range. Prices fell to such low levels that the prices farmers received for cattle shipped to market barely covered transportation costs.

On June 1 the government began an extraordinary cattle purchase program in the drought areas. By June 19, when special funds were granted by Congress for drought relief, the AAA had already purchased 213,124 head of cattle in North Dakota, South Dakota, Minnesota, and Wisconsin. County agents were again looked to for leadership and served as the directors of the cattle purchase program. Each agent appointed a committee of three men from each county to determine the price the government would pay for the cattle. Representatives of the Bureau of Animal Husbandry and the Department of Agriculture, as well as locally accredited veterinarians, assisted in inspecting the animals. Cattle were bought only in counties designated for emergency drought relief by the Department of Agriculture. Cattle over two years of age were purchased for from $6 to $14 a head, those under two years for from $5 to $10 a head, and those under one year of age for from $1 to $5 a head. In addition to the regular purchase price, a service payment, which was not subject to any mortgages that might be held on the cattle, was also paid to farmers and ranchers selling their cattle to the government. These payments were made at a flat rate of $6 for cattle over two years of age, $5 for those from one to two years old, and $3 for those under one year of age. Also, in spite of complaints that the government was trying to ''regiment'' farmers and ranchers, the purchase contracts provided that the seller had to agree to cooperate with the AAA in any future reduction program.

By the middle of July, as the drought spread to the corn belt, government officials designated emergency drought counties in Kansas, Minnesota, Missouri, Nebraska, North Dakota, South Dakota, Wisconsin, Illinois, Indiana, and Iowa. Pastures turned brown; waterholes dried up. In Missouri, nearly 30 percent of the farmers in the state were reported to be hauling water to keep their herds alive. Throughout the drought area cows broke down fences and wandered off in search of food and water. Farmers and ranchers surrendered to the sun during the day, but tied bells around the necks of their herds so they could be found at night and returned to their home pastures. Some farmers and ranchers still complained that the New Deal was dictatorial and that government prices were too low, but most saw the cattle purchase program as their only hope.

Cattle bought by the government were delivered to field agents of the Federal Surplus Relief Corporation (FSRC), which in turn shipped the cattle to be fattened or slaughtered. Cattle unfit for human consumption were immediately destroyed and buried on the farms

Farmers tried a variety of methods to control the grasshopper plague. Here a farmer uses a grasshopper net to clear his fields of the infestation. (Mildred Rothgarn Collection, State Historical Society of North Dakota)

and ranches where they were purchased. By July 26 the government had purchased 1,545,915 head of cattle, 176,000 of which were immediately condemned and 171,000 of which were shipped to Southeastern states for fattening before they were butchered. With cattle pouring in so fast, the FSRC was unable to process them as quickly as they were delivered. To meet the crisis, the FSRC contracted with privately owned feedlots to care for as many cattle as possible until they could be sent to processing plants. Privately owned canning plants handled most of the cattle, but a number of special canning factories, using relief labor, were also opened to handle the emergency.

By late summer the cattle purchase program took on the semblance of a wartime emergency as the government's "field armies" moved through the countryside. The government continued buying cattle through the fall, but by December purchases were nearly complete. Nationally, by January 2, 1935, the government had purchased 7,815,026 cattle from 675,499 farms in twenty-four states. In the Middle West the government purchased 3,686,887 cattle, at a cost of $51,472,734, in 540 emergency drought counties.

While cattle producers were aided by the government almost from the beginning of the drought, sheep and goat owners were not given any assistance until the middle of August when the AAA began purchasing sheep and goats in the drought areas. Nationally the AAA purchased 3,596,867 sheep and 354,432 goats in twenty states by December 1, 1934. For ewes one year old or over, the government paid $2 a head, for angora goats of the same age, $1.40 per head. One-half of the payment was in the form of a service fee, which like the service payment for cattle was not subject to mortgages. In the Middle West, 289,244 sheep and goats were purchased at a cost of $509,527.

The Department of Agriculture accomplished its major goals with the emergency slaughter campaign. The surplus of cattle, sheep, and goats was eliminated with sudden efficiency. Many animals were destroyed at the purchase sites, but millions of pounds of meat were saved and later given to poor people on relief. The administrative structure established by the government worked quickly and effectively to save many cattle producers from complete economic ruin. The quality of many big herds was improved as a result of the reduction program. The drought served as a culling process, because

FERA drought relief workers finish a farm dam in Adams County, Nebraska. (Nebraska State Historical Society)

Technically the "dust bowl" was limited to the area where Texas, Oklahoma, Kansas, and New Mexico join. Dust storms were, however, common throughout the Middle West. (Kansas State Historical Society)

little monetary difference was paid by the government for quality or poor stock; farmers naturally sold their worst cattle to the government. Small farmers and ranchers fared less well. Many farmers, particularly those who owned only small herds, found their foundation stock, which had been carefully built up over a long period of years, wiped out by the drought and the government's cattle-buying program. Poor farmers, who might have a single milk cow that the family needed for subsistence, were devastated by the loss of their livestock.

The AAA wanted to reduce the number of cattle on farms and ranches, but it also took measures to help farmers who did not want to sell their livestock. As feed became scarce in the drought areas, the AAA was forced to make modifications in crop adjustment contracts to enable farmers to grow forage crops on idle land. By the end of June the AAA made a series of rulings which lifted all restrictions on the planting of feed crops on all contracted and noncontracted acreage in the drought areas. The AAA also moved to conserve feed supplies and to manage their usage. Farmers and ranchers had very little cash, but hay, alfalfa, and corn prices skyrocketed. To help farmers locate feeds the AAA established a clearinghouse of information in Kansas

City, Missouri. Areas that had surplus feed were requested to notify the office if they wished to sell feed, or if they had pasture lands that could be rented to farmers whose pastures had burned up. The administration also moved to remove import restrictions on livestock feeds. On August 10 President Roosevelt, after completing a trans-continental trip across the drought area, signed a proclamation waiving import duties on Canadian hay and other forages. A new cooperative, the Agency for Deficiency Distribution, was created to import grains from Canada to increase livestock feed supplies in the drought areas. The imports, which were mostly frost-damaged wheat, were underwritten by the federal government.

Efforts to locate and redistribute feed supplies, coupled with the shipment of cattle to and from the drought areas, resulted in heavy railroad traffic. Consequently both state and federal officials worked to lower freight rates in the drought areas. In Washington an inter-departmental Committee on Transportation was created to coordinate the activities of the various state and governmental agencies concerned with freight rates. Railroad executives complained that they could not afford to reduce their rates, but they ultimately capitulated to pressure applied by the government and the various farm organizations. On June 4 the railroads reduced rates in the drought areas to two-thirds of normal on feed, 50 percent on hay, straw, and water, and 85 percent on cattle shipments from the drought area, with the privilege of returning the cattle to their home pastures at a later date for 15 percent of the normal rate. By October 1, 1934, the Interstate Commerce Commission estimated that rate reduction had already saved farmers in the drought areas an estimated $5,838,000. Farmers realized, throughout the summer of 1934, that government assistance was indispensable as they fought to survive the combined effects of drought and depression. Agrarian radicals still urged farmers to organize and warned that the administration was creating a system of "lords and barons," with farmers and laboring people as "feudal serfs," but the charge increasingly had a hollow ring.

The drought legitimized the need for planning and for an administrative structure in Washington that could deliver immediate relief to the farm community. Ultimately the drought enabled Roosevelt to co-opt farmers into the New Deal's administrative framework by providing benefits and services farmers needed to survive. The drought cut the heart out of the radicals' contention that the New Deal

This dust storm in Iowa in 1933 closed highways and blocked out the sun. (State Historical Society of Iowa)

served only the rich and the powerful. The temper, if not the philosophy, of farmers changed. By the end of the summer Senator Capper observed of farmers in his own state, ''There is not a trace of revolutionary spirit among them.'' The same was true throughout the drought-stricken states.

The Federal Emergency Relief Administration also continued to provide valuable support to farmers throughout the country. During its first year of operation, more than 10 percent of the FERA's clients were in rural areas. In the Middle West the number of farmers on relief varied, but it was not uncommon to find nearly a third of the farmers on relief. In the drought areas, relief rolls, particularly on the western fringe of the Middle West where highly speculative ''suit-case'' farming was common, were much higher. Many farmers left the land in 1934; for those who stayed, cash grants, food distribution, and employment on FERA work projects were necessary for sur-vival. By early 1934, before the drought became severe, there were already more than six hundred thousand farm families on relief. President Roosevelt realized that emergency relief programs were necessary to keep farmers alive, but he was anxious to make farmers self-supporting again as soon as possible. On February 28 he announced that in the future the government's rural relief program would be distinguished from urban relief. Roosevelt believed that farmers could be made self-sufficient, and permanently removed from relief rolls, if they were provided with sufficient credit, stock, land, and equipment and were taught how to live in harmony with the land.

The contradiction of developing plans to increase the productivity of poor farmers while, at the same time, the AAA worked to find ways to cut back agricultural production, seems not to have worried the President. Roosevelt still believed that small, poor farmers could be saved by encouraging more efficiency in their farming operations. On March 22 Harry Hopkins announced that all direct relief and Civil Works Administration projects in rural areas would be replaced by a national program of "rural rehabilitation." To administer the program, a new unit, the Division of Rural Rehabilitation and Stranded Populations, under the direction of Col. Lawrence Westbrook, was established in the FERA.

Ultimately three basic programs, rehabilitation, resettlement, and land-use planning, were developed by FERA officials to aid the rural poor. The most important, rehabilitation, involved making loans or grants to farmers on good land to purchase seed, fertilizer, livestock, and equipment. For those farmers who could not become self-supporting on their own land, the FERA planned, after purchasing their farms, to resettle them on more profitable land or in cooperative communities established by the government. All farmers would be taught better land-use management techniques.

In 1934 the FERA still concentrated most of its attention on providing direct relief to farmers. On May 19 Hopkins promised that federal expenditures for work projects and direct relief would be increased by $6 million per month until the drought was broken. The Division of Rural Rehabilitation also made a modest effort to begin "rehabilitating" farmers. By December 1934 the FERA had made rehabilitation advances to 68,625 farmers, 7,595 of whom were located in the Middle West. The FERA planned to retire 100 million acres of land from production and to resettle thousands of farmers,

Cowbells became hot-selling items in the 1930s. During the worst storms cows broke down fences and wandered far from home. The bells helped farmers find their livestock after the storms had subsided. Many animals, however, suffocated and died on the range. (Kansas State Historical Society)

but the land purchase program met with stubborn resistance and made little progress during the year. Farmers in the Middle West wanted relief, not resettlement. They were more responsive to new land-use planning programs, but the drought guaranteed that the FERA's major focus in 1934 would continue to be emergency relief.

By the middle of July the FERA was caring for four hundred thousand families in the drought areas; by the end of the month it was caring for eight hundred thousand families, half of which were receiving direct relief and the remainder of which were employed on work relief programs. Julius Stone, who represented Hopkins in the Middle West, emphasized that the work projects in the drought areas should have long-range objectives and should include the construction of more farm and garden ponds, the conservation of water, and the retirement of submarginal lands. Farmers in the Middle West were interested in the government's conservation plans, but were more worried that their families might starve to death. On August 14 Roosevelt conferred with Secretary Wallace, and Aubrey Williams, the deputy director of the FERA, and was informed that, since the Congressional appropriation on June 19, the drought area had increased by 300 percent. Again, the works program was speeded up in the drought area. In September the Bureau of Public Roads announced that it would also provide work for drought-stricken farmers on public roads.

During the winter months much of the work on construction projects had to be suspended because of the weather, but the government still provided farmers with cash relief payments. From October 1934 to February 1935, the number of farmers needing relief rose by nearly 75 percent. During the same four months 685,000 farmers, nearly 10 percent of the national total, received general relief grants or rehabilitation loans from the government. In the spring wheat areas of the Middle West, which had been particularly hard hit by the drought, nearly half of all farm operators were on relief by the end of the year.

The Farm Credit Administration also expanded its credit operations in the drought areas. On February 23, 1934, the FCA was given $40 million to make loans for crop production, harvesting, and summer fallowing, and for the purchase of feed for livestock. These loans, which required a first lien on crops or livestock as security and charged 5.5 percent interest, were given to farmers throughout the nation but were especially helpful to farmers in the drought areas. Originally the maximum loan available under the program was $250; by July, as the drought heightened in intensity, the amount of the loans in the drought areas was increased to $750.

Another $96,785,900 was given to the FCA for emergency crop and feed loans from the $525,000,000 drought appropriation. Dudley Doolittle, general agent of the Tenth Bank District of the FCA, assured farmers that the loans would be made on a relief, rather than a commercial, basis, and promised farmers that the government would ''not let him perish, nor his livestock die, for the lack of money to buy food for his family, and feed for his livestock.'' Led by South Dakota, North Dakota, Nebraska, Minnesota, and Kansas, 133,198 loans totaling $21,922,371 were made in the Middle West during 1934.

The FCA also continued to make Federal Land Bank and Land Bank Commissioner loans to farmers. In 1934, 267,958 Land Bank and Land Commissioner loans, totaling $765,500,849, were made to farmers in the Middle West. Almost all of the money was used to refinance farm mortgages. Increasingly farmers looked to the

Farmers watch an approaching dust storm. Many stayed inside during the worst storms, but "dust pneumonia" became a common ailment in the 1930s. (Kansas State Historical Society)

government, not private capital, for financing. Farmers whose crops had failed now feared that the government, rather than banks or insurance companies, would foreclose their property. William I. Myers, governor of the FCA, made it clear that government loans had been made with "every expectation" that the money would be repaid, but also promised that "no farmer making an honest effort to take care of his obligation to the Farm Credit Administration, who is unable to do so because of the drought or crops failure, need fear that we will close him out."

Of necessity, the government concentrated on emergency relief programs in 1934, but the drought also made it clear that a national land-conservation program was badly needed. Savage dust storms became increasingly frequent in the Great Plains. As the nation became more "erosion conscious," a new phrase, "the dust bowl," became a common topic of conversation throughout the country. Technically, the dust bowl was limited to southeastern Colorado, the Oklahoma Panhandle, the western third of Kansas, and most of the Texas Panhandle. Erosion was, however, extensive throughout the Middle West, and the entire region suffered from the effects of blowing dust. On May 3, 1934, a storm carried an estimated three hundred million tons of soil from the drought area eastward until it reached New York and drifted more than five hundred miles out to sea. As the dust rolled across the land doctors reported an increase of from 50 to 100 percent in the number of reported cases of ''dust'' pneumonia and sharp increases in the infant mortality rate. In Kansas alone doctors attributed at least three hundred deaths to the blowing dust and unbearable temperatures.

Before farmers went outside during the worst storms, they tried to protect their eyes and faces with goggles, wet hankerchiefs, or masks crudely fashioned out of gauze or old rags. In some areas the Red Cross distributed gas masks, which had been made during the First World War to protect soldiers in the trenches, to farm families so they could cover their children's faces at night to protect them from suffocation. Women, driven inside their homes for days at a time by the blowing dust, stuffed wet rags and sheets around cracks in doors and windows in a vain attempt to keep the dust from filtering into their homes. When the worst storms were over and farmers went outside to survey the damage, they found that the dust had buried fences, cars, machinery, sheds, trees, and even houses.

Some farmers blamed the storms on Roosevelt and insisted that the drought was God's punishment for killing ''little pigs'' and for plowing under crops while millions of Americans faced starvation. Other farmers were more stoic and grudgingly surrendered to the forces of nature. In South Dakota signs which read, ''If at first you don't succeed . . . The Hell with it,'' were left behind as farmers abandoned their homesteads. Those who remained to face still more storms developed a bitter and ironic sense of humor to cope with their individual tragedies. Women, who learned to turn their plates upside down to keep out the dust when they set their tables, told about the woman who cleaned her dishes, pots, and pans by putting them out the window to be blasted clean by the blowing sand. Farmers laughed about the rancher who went to the bank to secure a loan and looked up to see his farm blowing past the window. Others told the story of a rancher who bought a bucket of gravel to throw on his rooftop at night so that his children would know the sound of falling rain; or about frogs who drowned when they were thrown into a pond because they had never learned to swim. The story was also told about the man who fainted when he was hit by a drop of rain and had to be revived by throwing a bucket of dirt in his face. Farmers insisted that they

An important part of President Roosevelt's drought relief program in 1936 was the Great Plains Shelterbelt project. Here workers plant trees in Kansas. (Kansas State Historical Society)

gauged the intensity of the great storms by tying a log chain to a tree: if the chain blew straight, the wind was calm; if it popped like a whip there was a breeze; if it uprooted the tree, there was a blizzard. Farmers were less than amused, however, by the charge that they were largely responsible for the ecological disaster which caused so much of their suffering.

The most obvious and immediate cause of dust storms in the 1930s was the drought. Great Plains farmers pointed out that dust storms were not new; their forebears in the nineteenth century had also experienced savage dust storms during dry-weather cycles. Government officials concluded, however, that the cause of the storms was not just drought, but unsound farming practices. Henry Wallace and Harry Hopkins warned that the nation's greatest natural resource, the land, was being destroyed by wind and water erosion. Wallace did not accuse farmers of being greedy when they moved onto marginal land, but he did conclude that millions of acres of damaged farm land in the Middle West would have to be taken out of production. Wallace lectured farmers that the damage done to the land by the drought would be healed by rain, but the damage ''done our land by generations of haphazard, misplaced settlement, overcropping, exploitation and permitted erosion will never heal unless we take hold of the situation, and keep hold, with a long-time program of soil repair, settlement, and balanced harvests.''

Before the 1930s the government's record in providing farmers with useful information about soil-conservation practices was poor. The vision of the West as an unlimited and inexhaustible resource was more frequently than not shared by the Department of Agriculture. By the 1920s many farmers had begun to develop their own soil-conservation techniques, but the United States remained the only major country in the world without a national policy to govern the use of the nation's land and water resources.

When Roosevelt became President he was determined to promote better land-use practices. In 1933 the Civilian Conservation Corps (CCC) was created to provide work for unemployed young men, primarily from urban areas, on projects that would restore or protect the nation's long-neglected landscape. In 1934 the CCC was also used to provide relief in the drought areas. In July the CCC issued a special call for fifty thousand enrollees from the drought states in the Middle West. The drought camps worked on a number of projects, including flood control and reforestation, but they were primarily concerned with projects that would control soil erosion. The Department of the Interior, through its Soil Erosion Service, also sponsored demonstration projects in the drought areas to teach farmers conservation techniques to protect the dry, barren land from further wind and water erosion.

To speed the development of a national land policy, Roosevelt created, on June 30, 1934, the National Resources Board. The Board, which was chaired by the Secretary of the Interior, Harold Ickes, was charged by the President with the responsibility for studying methods of restoring land that had been damaged through years of neglect, and with formulating plans to move farm families away from stricken land to more productive agricultural regions. The Board was to report its findings by December 1.

While the National Resources Board began conducting surveys, Roosevelt began another program which also promised long-range relief to farmers in the Middle West. On July 11 he set aside $15 million from the $525 million Congressional drought relief appropriation to begin the Great Plains Shelterbelt Project. The project called for the planting of a hundred-mile strip of trees along a line extending through the Dakotas, Nebraska, Kansas, Oklahoma, and Texas. It was estimated that the project, which might take ten years and $75 million to complete, would affect nearly twenty million acres of land.

The announcement of the project, as might have been expected, was followed by sensational publicity in the press. Because the project had been initiated to cope with the drought, many people inferred that it would change the climate, stop high winds, and create ideal farming conditions in the areas surrounding the shelterbelt. The *New York Times* stated that hundreds of thousands of tons of water would be saved by the tree foilage and that the wind velocity would be slowed by 35 percent in the summer and 20 percent in the winter. F. A. Silcox, chief of the Forest Service, tried to stem the tide of speculation by emphasizing the immediate benefits of the program for the drought areas. One estimate stated that 520,000 man-hours of work would be needed during the first six months of the project, and ultimately 190,000 pieces of land and thirty to fifty million fence posts would be needed to complete the shelterbelt and to build fences to protect the young trees. In addition to stressing the project's immediate impact, Silcox pointed out that such long-range benefits

Young trees planted to create shelterbelts to control wind erosion. (Kansas State Historical Society)

as supplies of wood and timber and decreased soil erosion could be expected from the shelterbelt in future years. He made no grandiose promises, however, that the project would change the climate or remove the possibility of future droughts.

Another farsighted measure sponsored by the AAA was the seed-conservation program. As the drought took its toll on crop and feed supplies, a danger developed that seed, which had been carefully adapted after years of research and experimentation, might be destroyed or would not be available in sufficient quantities for planting the following year's crop. Consequently, a Seed Conservation Committee was established to make a survey of seed supplies and to formulate a plan to insure that all superior seed would be conserved. To finance the program, on June 23 Roosevelt allocated $25 million to the AAA from the $525 million drought relief appropriation. On the same date a ruling from the AAA allowed farmers to harvest seeds from pasture and meadow crops that were under contract to the government.

The drought helped the government win support for long-range conservation programs, but by late summer the drought also provided opponents of the AAA's production control program with new ammunition to launch an attack against the New Deal's philosophy of scarcity economics. Before the drought was temporarily broken in the fall, the wheat crop had been reduced to one-half of the yearly average; corn production was the smallest in thirty-four years; oats, rye, and buckwheat production was the smallest in half a century. Many farmers harvested no crops at all. AAA officials found themselves in the incongruous position of supporting a program designed to ''artifically'' reduce agricultural production while the drought ''naturally'' destroyed crops and eliminated surpluses that had glutted the market.

Pressure to discard the AAA's production control programs poured in from every section of the country. From the East, as food prices rose sharply, people complained that the Administration was favoring the Middle West at the expense of the industrial sector of the economy. The drought, combined with the Administration's control programs, also caused a momentary panic that there would be a national food shortage. Farmers in the Middle West, temporarily freed from the dread of plenty, also questioned the wisdom of continuing production controls. Many still believed that the drought was God's way of punishing the Administration for tampering with the "laws of nature."

Wallace pleaded with the East to have patience. The Secretary insisted that the real danger was that farmers, with the return of good weather, would increase production, causing farm prices to spiral downward again. Declining farm prices, Wallace warned, would drag down the entire economy and would ruin the Administration's recovery program. Wallace lamented the fact that people seemed to think that pigs were raised for pets and were more concerned about their welfare than the well-being of the farmers. Angrily he charged that those who believed the drought was a judgment from Heaven were agents of "chaos" who wanted to protect the "old order." He was also careful to point out, however, that the dry weather could not be the result of God's anger with the AAA because Canada, Poland, Russia, and China were also suffering from drought.

New Deal opponents who wanted to return to unlimited agricultural production counterattacked by charging that Wallace was personally benefiting from the farm crisis and that he was guilty of a conflict of interest. For years Wallace had been interested in creating a new corn hybrid which would increase agricultural yields and had held a controlling interest in the Hi-Bred Corn Company of Grimes, Iowa. Wallace resigned as director of the company before he became Secretary of Agriculture, but his wife continued to hold stock in the company. The Chicago *Tribune*, one of the leading conservative newspapers in the country, uncovered the stock arrangement and quickly charged that Wallace was guilty of gross misconduct. The newspaper concluded, "Mr. Wallace has profited from the misery which he has helped to create. . . . Reduce your corn acreage by 20 percent, as the government requests, says the Secretary of Agriculture, and you can collect the bounty which the government is paying to farmers who do so. Then plant our seed, at $7.00 per bushel, says Mr. Wallace's company, and you can grow as much corn as you formerly did."

The irony of trying to persuade farmers to limit production, while at the same time his family's company sold products that enabled farmers to produce even more, could not have escaped the Secretary. His dilemma symbolized the problem of the age. Wallace wanted farmers to become more modern and efficient by taking advantage of new hybrid seeds. He also realized there were no markets for the new

Women stuffed rags and wet sheets around windows and doors during the dust storms, but dust filtered in anyway. (Kansas State Historical Society)

abundance. Lamely Wallace pointed out that the Hi-Bred Company produced only enough seed corn, after many years of research and experimentation, to seed less than 1.5 percent of the corn acreage planted each year in Iowa. He could not, however, rationalize the irrational: want in the midst of plenty, increasing productivity at a time when the farmers' productive capacity had already reached suicidal dimensions.

The Farmers' Union and the Farmers' Holiday Association joined in the attack. The Farmers' Union lost its most eloquent spokesman when John A. Simpson died, in February 1934, but other Farmers' Union spokesmen, including C. N. Rogers, continued to verbally assault the "Brain Trust" and insisted there would be a national food shortage within six months. Rogers suggested that if Roosevelt forced Wallace and Tugwell to live on a farm for a few months they would rapidly change their policies and might finally do something for agriculture. Milo Reno continued his harangues against Wallace, "the worst enemy agriculture ever had," and concluded that rather than trust the Secretary's judgment, "I would prefer to entrust to a crap game of loaded dice." Rogers and Reno insisted that farmers were still opposed to the domestic allotment plan, and Reno threatened another strike; but it was obvious by the summer that he had

little support in the farm belt. On September 21, 1934, the Holiday
Association voted against a farm strike, explaining that since farmers
had nothing to sell, a strike would be futile.

A drought of serious proportions in 1933 might have aided left-
wing farm organizations in their battle against the domestic allotment
plan. In 1934, however, although the AAA remained unpopular, the
massive relief efforts of the government in the drought areas won the
support of the vast majority of Middle Western farmers. Roosevelt
was more popular than ever. Ole Nordland, editor of the Cedar
Rapids (Nebraska) *Outlook*, informed the President, ''You are indeed the idol of the farmers for through a season of adversities, such
as they have faced, they yet uphold you and support your every effort
and plan.'' Wallace remained controversial, but after a period of
momentary indecision the Secretary and officials in the AAA soon
found compelling arguments, in spite of the drought, for continuing
production controls.

The most effective argument used by defenders of the AAA was
that government benefit payments served as a form of ''crop insurance'' during dry weather. George E. Farrell, Chief of the Wheat
Section of the AAA, explained that the wheat contracts guaranteed
farmers an income in years of drought as well as years of surplus.
When it became apparent that 1934 would be a drought year, the
AAA even reopened the time for signing adjustment contracts to
enable more farmers to receive benefit payments. Rexford Tugwell,
then Assistant Secretary of Agriculture, added that the AAA's
program was not necessarily one of reduction; taking the drought and
the international situation into consideration, it might even encourage
full production. AAA officials also argued that acreage reduction in
times of drought was actually a positive benefit, since farmers saved
the cost of seed and labor on crops that would have been abandoned
anyway. Wheat farmers were supposed to plant 54 percent of their
base acreage to crops to be eligible for benefit payments in 1934.
When it became obvious that such planting would be a wasted effort,
the AAA modified its wheat adjustment contracts in Kansas, Minnesota, Nebraska, North Dakota, and South Dakota to allow farmers
to abandon the land and still receive payments from the government.

AAA officials were also quick to point out that the drought, not
production control programs, caused the critical feed shortage in
1934. The AAA estimated that nationally 36,767,000 acres of land

were taken out of production, but by the end of the year only 5.4 percent of the land was still idle. New seedings of alfalfa, grass, and other emergency forage crops actually increased the acreage used to grow feed for livestock. In the corn belt states of Ohio, Missouri, Indiana, and Iowa, pasture acreage increased from 37,779,340 acres in 1929 to 42,787,452 acres in 1934.

Officials in Washington also launched a major campaign to persuade consumers that the short harvests in 1934, and the dramatic increase in farm prices, were due to the drought, not the AAA's production control program. The wheat crop was the smallest in forty years. The acreage seeded to wheat in 1934 totaled 60,371,000 acres, enough to produce in a normal year 750,000,000 bushels of wheat. Contract signers reduced their acreage in 1934 by an estimated 23 percent, but noncontracting farmers increased their production so that the actual acreage planted was only 8.5 percent less than the 1930-1932 average. The average yield in 1934, however, was only 8.5 bushels per acre, compared with the normal yield of 12.5 bushels per acre. Consequently, wheat production for the 1934-1935 season totaled only 496,929,000 bushels, compared with the 1928-1932 average of 860,570,000. AAA officials estimated that the drought had reduced production by 309,417,000 bushels, while the AAA's program had reduced production by only 54,224,000 bushels.

The average price of wheat in 1934 rose to 84.7 cents per bushel. Farm income from wheat production increased in 1934 to $289,169,000; benefit payments totaled $101,600,000, giving wheat farmers a cash income of $390,769,000. Farmers in the drought areas, however, saw their income fall to even lower levels. In two major wheat-producing states, North and South Dakota, farm income was less in 1934 than in 1933. In South Dakota, where benefit payments from the government exceeded $27 million, farm income fell from $68 million in 1933 to only $55 million in 1934. In North Dakota benefit payments totaled $37.5 million, but farm income still fell from $75 million in 1933 to only $63 million in 1934.

The drought also reduced corn and hog production much more than the AAA had planned. The AAA reduced corn acreage in 1934, anticipating that the final crop would total 2.2 billion bushels. Actual production was only 1,377,126,000 bushels, the smallest crop since 1880. AAA officials estimated that the control program had reduced production by 181,685,000 bushels, and that the drought had reduced the crop by 1,003,336,000. The AAA had planned to reduce hog production in 1934 by thirteen million litters; the actual reduction, as a result of the drought, was nearly double that amount. The limited supply of corn and hogs resulted in substantial price increases. Hogs sold for an average of $4.17 per hundredweight in 1934, and corn for an average of 64.5 cents per bushel.

The corn loan program continued in 1934, but, with rising prices, the number of applications fell dramatically. In 1934 the Commodity Credit Corporation loaned $11,941,457 on 20,073,395 bushels of corn to 15,697 farmers. By the end of the year, with corn selling for eighty-five cents a bushel, most of the loans were repaid. By December 31, $119,949,367 of the $120,493,034 loaned in 1933 had been repaid to the government.

Many farmers, especially in the drought areas, had managed to survive in 1934 because of the benefit payments they received from the government. Still, with prices rising, AAA officials feared that farmers would now turn their backs on production controls. During the first two weeks of October a referendum was held among corn-hog producers to determine whether they wanted to continue the adjustment program in 1935. Less than 50 percent of the farmers eligible to participate in the referendum actually voted. Of those farmers who had signed adjustment contracts in 1934, 69.9 percent favored continuing the program; of the nonsigners who voted, only 33.1 percent were for continuing the AAA's corn-hog reduction program. The vast majority of farmers voted in favor of the program, but in key states such as Kansas and Nebraska less than 50 percent of the voters wanted to continue production controls.

While the Administration claimed victory, it was apparent that many farmers voted for the AAA not because they accepted the need for planning but because they still needed cash subsidies from the government. President Hoover, obviously enjoying his successor's problems, cynically wrote to his former Secretary of Agriculture, Arthur Hyde, "The astonishing thing is that there is any opposition. I should have thought, that Wallace would have taken a vote on abolishing Santa Claus." Ultimately, the AAA decided to continue its program of production controls, but it liberalized corn-hog contracts for the following year. The hog allotment was to be 90 percent of the base period, and farmers were allowed to plant up to 90 percent of their base acreage to corn in 1935.

The AAA's dairy program continued to experience difficulties. On February 1, 1934, the AAA terminated the fluid milk marketing agreements instituted in 1933 and implemented a new policy. New marketing agreements, which fixed the prices producers received for their milk, were issued, but the AAA abandoned its previous efforts to fix distributor prices. Since surplus milk could still be processed as butter or cheese, the new marketing agreements failed to stabilize prices for the dairy industry as a whole. Wallace tried to persuade dairy producers to support production controls, but the National Conference of Dairy Co-operatives went on record as opposing "governmental action aimed at placing in the hands of the Federal Government complete and bureaucratic control of the fundamental activities of our people, including agriculture, industry and finance." The dairy cooperatives recommended dropping the AAA and reviving the stabilization operations of the Federal Farm Board. The drought, by limiting pasture and feeds, did force the liquidation of a number of dairy herds and reduced production, which in 1934 was 3.3 percent less than in 1933. Dairy prices, in spite of the absence of production controls, rose to 82 percent of parity; farm income from dairy products also increased, to slightly more than $1 billion.

Reluctantly, the AAA continued its surplus removal program for dairy products in 1934. Between January 2 and April 4 the AAA purchased 6,346,265 pounds of cheese for distribution to relief clients. Between July 2 and December 8, 10,135,960 pounds of butter and 367,000 pounds of cheese were purchased. By the end of the year the AAA had commitments to purchase 67,748 pounds of butter and 17,803,224 pounds of cheese.

It had been an incredibly difficult year for farmers, but the end of 1934 brought a general improvement in the agricultural situation. Farm income in the United States rose to $6 billion, an increase of nearly 20 percent over farm income in 1933. AAA officials estimated that 7.3 percent of the farmers' income, or $466 million, came in the form of rental and benefit payments from the government. The average price of farm products, which had been 77 percent of the prewar value in January, rose to 101 percent of the prewar value by September. Prices that farmers had to pay, however, had increased by 126 percent of their prewar value, making the actual value of farm products only 80 percent of parity. The volume of farm exports was 36 percent less than in 1933, but the value of farm exports increased from $500 million in 1932-1933 to $669 million in 1934-1935.

Throughout 1934 the government was forced to focus its attention and to commit its resources to emergency drought relief programs. The problem of surpluses was temporarily resolved by the drought, but Wallace still worried about "the fundamental absurdity that confronts us all, . . . the simultaneous existence of man power, technical facilities, and resources for enormous production of goods, and of millions of people in dire poverty, wanting and needing the goods." He agreed with William Hirth, president of the Missouri Farmers' Association, who wrote to Roosevelt, "That we should have idle and hungry and ill clad millions on the one hand, and so much food and wool and cotton upon the other that we don't know what to do with it, this is an utterly idiotic situation, and one which makes a laughing stock of our genius as a people."

Wallace was still stung by charges that the Administration's plan to create artificial scarcity through production controls neglected the needs of the poor. He pointed out that even with the drought, agricultural production in 1934 was only down by 15 percent while industrial production was down to only 42 percent of predepression levels. A general improvement in the situation, Wallace advised the President, could only come with the recovery of the industrial sector of the economy. Wallace, and the Department of Agriculture, remained committed to production controls. They were still haunted, however, by want in the midst of plenty and by the continued existence of widespread poverty in both urban and rural America.

Chapter 5

New Approaches
to Rural Relief

In 1933 and 1934 the federal government had fought a successful holding action against the combined forces of depression, radical protest, and drought in the Middle West. Farm prices had risen, but the back of the depression still had not been broken. An impressive number of programs to bring relief to farmers had been put in place, but the problems of the rural poor remained substantially unchanged. Soil erosion and the abuse of the land had been identified as a major concern, but the Administration's soil-conservation program was still in its infancy. The AAA officials remained optimistic, but they were apprehensive about the apparent lack of enthusiasm among Middle Western farmers for the Administration's farm program.

In 1934 the Democrats again won a smashing political victory in the Middle West. In the Congressional elections important Republican critics of the New Deal, such as Arthur R. Robinson of Indiana, Roscoe C. Patterson of Missouri, Simeon D. Fess of Ohio, Robert Simmons of Nebraska, and Harold McGugin of Kansas, all met with defeat. In gubernatorial contests, the only Republican victory came in Kansas, where Alfred M. Landon swept to an easy victory. There were already signs, however, that many farmers were returning to the Republican Party. In North Dakota the Democratic vote fell by 16 percent; in South Dakota by 7 percent; in Nebraska by 8 percent; and in Iowa, by 5 percent.

Roosevelt's opponents on the left were also encouraged by the elections. In Minnesota Governor Floyd B. Olson's Farmer-Labor Party captured 49 percent of the vote to only 30 percent for the Democrats; in Wisconsin the Progressive Party, which was revived by Robert and Philip La Follette, captured 50 percent of the vote to

Rexford Tugwell, who headed the Resettlement Administration, meeting with Ross Smith, August 1936. (Franklin D. Roosevelt Library)

only 26 percent by the Democrats. The success of Olson and the La Follettes encouraged radicals to believe that a national third party would attract widespread support in the Middle West.

In July 1935 a new organization, the American Commonwealth Political Federation, was formed in Chicago to work for the formation of a new third party. The Federation, which was dominated by humanitarian socialists and left-wing intellectuals such as Thomas Amlie of Wisconsin, Paul Douglas of the University of Chicago, the Rev. Howard Y. Williams of St. Paul, and Alfred Bingham, Nathan Fine, and Selden Rodman of New York, developed a platform that promised farmers cost-of-production, mortgage refinancing at 1.5 percent interest, and tariff protection for the farmers' domestic markets. Federation leaders hoped to unite labor and agriculture to challenge Roosevelt at the polls in 1936. But, of the farm organizations, only the rapidly fading Farmers' Holiday Association paid any attention to the movement. The Federation also failed to find anyone to unite left-wing dissidents against Roosevelt. The La Follettes continued to talk about the exploitation of the poor by private monopoly, and Olson insisted that he was "radical as hell," but neither was willing to break openly with the New Deal or to allow their names to be used by the American Commonwealth Political Federation.

Father Coughlin, Louisiana Senator Huey P. Long, and Dr. Francis E. Townsend, an obscure physician in California, also began in late 1934 and early 1935 to apply pressure on Roosevelt to move leftward and flirted with the idea of forming third parties. Each depended heavily upon the farm vote for support. By 1934 Father Coughlin was a well-known public figure with a weekly radio audience numbering in the millions. Coughlin disliked the New Deal farm program from the beginning, but he did not break openly with Roosevelt until late 1934. On November 11 he announced the formation of the National Union for Social Justice. Charging that Roosevelt was controlled by international bankers and the United States Chamber of Commerce, Coughlin promised farmers inflation, the remonetization of silver, and cost-of-production farm prices.

Townsend also enjoyed widespread popularity in the Middle West. Like many others in the 1930s, he was appalled by the fact that there was widespread hunger, especially among old people, while the AAA complained about agricultural abundance. To end the depression, and to feed the hungry, Townsend proposed that every person over sixty-five years of age be given a monthly bonus of $200. Townsend Clubs sprang up everywhere. The "old people's crusade" appeared to be especially threatening in the Middle West because the club meetings often combined radical economic ideas with evangelical Protestant values. Old people, in a world marked by uncer-

"*Black Blizzard,*" *Inavale, Nebraska, March 15, 1935. (The Land Its People Museum. Red Cloud, Nebraska)*

tainty and change, frequently seemed as concerned about main-taining traditional morality as they were about the collapse of the economic order. In emotionally charged meetings Townsendites demanded not only that old people be given a guaranteed monthly income, but also that something be done about necking, women wearing lipstick, cigarette-smoking, and other signs of "moral decay."

Huey Long appeared to be Roosevelt's most formidable opponent on the left. Long's farm program was vague, but he supported cost-of-production and inflation and openly appealed to the more radical farm organizations for support. Still viewed as a Southern regionalist, Long was anxious to demonstrate his popularity outside the South. Like Townsend, Long was quick to exploit the contradiction of want in the midst of plenty. By 1935 his ''Share the Wealth'' program, which would have taken money from the rich to give to the poor, was hotly debated in the Middle West. Early in 1935 Milo Reno, still hoping to unite dissident left-wing elements into a cohesive third party, invited Long, Coughlin, Townsend, and Olson to address the third national convention of the Farmers' Holiday Association in Des Moines. Only Long accepted the invitation. Amid frequent shouts of ''amen,'' the ''Louisiana Kingfish,'' while munching peanuts, launched a scurrilous attack on the AAA and shouted that Wallace and his cohorts should be hung. After the speech a confident Long concluded, ''You know that was one of the easiest audiences I ever won over. I could take this state like a whirlwind.''

Roosevelt, looking to the 1936 elections, was obviously concerned by reports that the political situation in the Middle West was rapidly deteriorating. At the request of James Farley, one of the President's most trusted political advisors, Emil Hurja, of the Democratic National Committee, conducted a poll to explore the support that a third party candidate might have in the 1936 election. The survey showed that the ''demagogues,'' as Roosevelt successfully labeled his opponents, had a considerable following in the farm belt. In a Presidential election Hurja concluded that Long might draw, with a strong showing in the Middle West, as many as 2,750,000 votes. Hurja worried that in a close election Long might be able to draw enough votes away from Roosevelt to elect a Republican to the Presidency. But Long was soon shot to death by an assassin, removing Roosevelt's most serious challenger. His successor to the Share the Wealth movement, Gerald L. K. Smith, along with Coughlin, Townsend, and agrarian radicals such as Milo Reno and William Lemke, still worked to form a third party to drive Roosevelt from the White House.

Roosevelt broadened the New Deal's farm program in 1935 by creating the Resettlement Administration (RA), which was charged with the responsibility of improving the plight of the rural poor, and the Soil Conservation Service (SCS), which would work to save the land from abuse and general mismanagement. Roosevelt was anxious

In the late 1920s farmers expanded their farming operations into marginal areas that had never been put to the plow before. When the rains stopped in the 1930s, the result, as this desolate farm shows, was devastating for the land and the people. (State Historical Society of North Dakota)

Still another storm darkens the sky and fills the air with dust. (State Historical Society of North Dakota)

to undermine his opponents on the left and undoubtedly weighed the political advantages of creating the RA and the SCS. Both agencies were, however, logical extensions of the New Deal's prior commitments to provide relief to the needy and to develop a sound land-use policy.

At the same time that Roosevelt moved to expand the New Deal's reform program, he also allowed Chester Davis, who had succeeded George Peek as administrator of the AAA, to ''reorganize'' the AAA. Davis was more tolerant than Peek of Jerome Frank and other urban liberals in the AAA, but like his predecessor he did not intend to use the AAA as an instrument of social reform. The immediate incident triggering the ''purge'' concerned Black sharecroppers in the South, but the reorganization had a major impact on every agricultural region, including the Middle West.

In the South, and in some sections of the Middle West, it was apparent during the AAA's first full year of operation that many tenant farmers were being victimized by the AAA's production control programs. In some cases landlords, realizing that the benefit payments they would receive from the government were more than the income they received from their tenants, simply evicted the

The year 1935 brought the worst dust storms in the nation's history. The storms made the nation more conservation-conscious and persuaded many farmers to develop new farming techniques to save the soil. (Kansas State Historical Society)

tenants from the land. In other case, landlords used the threat of eviction to renegotiate contracts, requiring that tenants pay more rent or that they give the owners a larger share of their crops. The AAA ruled that tenant farmers were also eligible for benefit payments. Pressured by the AAA's legal staff, landlords in the South also ultimately agreed not to reduce the number of tenants on their farms when they retired land from production. There were, however, widespread violations of the agreement in both the South and the Middle West. In late January 1935 a number of tenant farmers in Elkhart, Kansas, held a public meeting to dramatize their plight. They wrote in protest to Secretary Wallace: "The ultimate results will be that the allotment, instead of being a benefit to tenant wheat farmers, will prove to be the cause of them being either dispossessed or having their rentals raised to such a point that absolutely no benefit is derived by them from the allotment program."

Early in 1935, while Davis was away on a trip, Frank decided to act on complaints that landlords in the South were violating AAA contracts. He issued an interpretation of the cotton adjustment contracts that required not only that landlords not reduce the number of tenants on their farms, but also that they keep the tenants in the same houses and on the same land. The ruling was intended to protect tenants from arbitrary landlords and to guarantee poor farmers a place on the land. Davis, however, with the approval of Secretary Wallace, immediately fired Frank and a number of his supporters in the Legal Division. He also announced, as he continued cleaning house, that the head of the Consumers' Counsel Division, Frederic Howe, would be replaced by Calvin B. Hoover, and that Gardiner Jackson, one of the most outspoken "radicals" in the AAA, had been dismissed from his position.

Frank's insubordination probably warranted his dismissal. Poor farmers had, however, lost an important spokesman within the Administration. Frank had battled the processors for higher prices and had insisted that the Administration do something to protect poor tenant farmers. Rexford Tugwell, who believed that Secretary Wallace was being pushed around by some "pretty sinister forces," interpreted Frank's dismissal as a major victory for reactionary farm leaders who were owned "body and soul" by the processors. Frank's desire to save the small family farmer from extinction appealed to Jeffersonian romantics, but the policies he championed within the AAA would have done little to change the status of the rural poor. Higher farm prices, and the benefit payments offered by the govern-

A Kansas farmer surveys the damage following a 1935 dust storm. (Kansas State Historical Society)

ment, would not have enabled poor farmers to escape their poverty.
Modernizing their operations would have resulted in even more
overproduction and would have decreased the need for farm labor,
threatening to depress farm prices once again. The conundrum of
how to save the rural poor was inherited, ironically, by another urban
liberal, Rexford Tugwell.

In his annual message to Congress on January 14, 1935, Roosevelt
conceded that "in spite of our efforts and in spite of our talk, we have
not weeded out the overprivileged and we have not effectively lifted
up the underprivileged." In the spring of 1935 government officials
estimated that 900,000 farmers in the United States earned less than
$400 a year; 500,000 lived on 100,000,000 acres of submarginal
land; and many thousands, in spite of the Farm Credit Administration, were still heavily in debt. To mount a direct attack on rural
poverty Roosevelt created the Resettlement Administration and appointed Tugwell administrator.

A number of federal agencies, including the Rural Rehabilitation
Division of the FERA, the Land Policy Section of the AAA, the Soil
Erosion Service, the Natural Resources Board, and the Subsistence
Homesteads Division of the Department of the Interior, had developed programs to address the problems of land use and the
problems of the rural poor in a haphazard way between 1933 and
1935. Inadequate funding, poor planning, and the lack of coordi-
nation doomed their efforts to failure.

Tugwell had argued for years that the problems of the rural poor
were deeply rooted in the nation's long history of land abuse. Years
of neglect had resulted in the physical depletion of millions of acres of
land; poor farmers who lived on worn-out land had little choice but to
rely upon credit or government relief to provide their families with
the barest necessities of life. Initially Tugwell hoped to make land
reform the most important work of the RA. Soon after he was
appointed to head the RA, Tugwell announced plans to retire nineteen million acres of submarginal land from production, to initiate a
program of land-use planning, and to move farmers who were living
on land purchased by the government to new locations where they
would have a reasonable prospect of making a living. In the Middle
West, the RA planned to purchase large tracts of land from two
particular areas, the Great Lakes cutover region and the western edge
of the Great Plains, which seemed to be turning into a "great

Dust piled up around a snow fence after a dust storm. (Kansas State Historical Society)

American desert.'' The land purchased by the government would then be turned into parks, recreational facilities, and refuges for migratory waterfowl. By December 1935, 5,032 men were in the field developing the RA's land-use program; by April 1936 the number increased to 57,751.

The Rural Resettlement Division of the RA sponsored two kinds of projects to relocate farmers who were displaced by the government's land purchase program. Under the ''infiltration projects,'' farmers were to be resettled on individual farms scattered throughout already established farming districts. Under the ''group settlement'' projects, a number of farmers were to be placed on adjoining farms and were to share certain community services, such as schools, canneries, and farm equipment provided by the government. Farmers were also settled in ''rural-industrial'' communities, which were designed to provide part-time work in both industry and agriculture. Resettlement officials argued that placing poor farmers on more productive land would not interfere with the efforts of the AAA to reduce agricultural production, because most of the land used for resettlement was already being used for agricultural production.

The RA's land purchase and resettlement program ran into immediate difficulties. Many farmers, even if they were poor, had no desire to leave the area where their families had lived for generations.

A Kansas farmer shovels dust after a storm. In the winter when blowing snow and dust combined, the storms were called "snusters." (Kansas State Historical Society)

Others, deeply suspicious of the motives and competence of officials in Washington, resented the government's meddling in their individual lives. Planning excited the imaginations of New Deal farm officials in Washington, but met with marked resistance in the countryside. From the beginning, the RA was also hampered by a lack of adequate funds to address a problem of such monumental dimensions. By June 30, 1936, after the RA had been in operation for nearly a year, the government held options on 9,365,617 acres of land but had actually purchased only 1,744,342 acres at a cost of $7,796,410. In the Middle West the government held options to purchase 2,682,214 acres of submarginal land and had purchased 524,706 acres at a cost of $3,413,769.

The promised resettlement of farmers also fell far short of Tugwell's expectations. By the end of its first year the RA was considering projects which, had they been carried out, would have resettled fifty-five thousand people on 664,530 acres of land in thirty-five states. The resettlement program, however, also stalled. In the Middle West only eight resettlement projects, containing 238 units and involving 1,394 acres of land, had been completed by June 30, 1936. The community program, which had been hailed as a "radical" solution to the problems of rural poverty, proved to be little more than an esoteric dream.

As an "emergency" measure, the RA also planned to provide "rehabilitation" programs for the rural poor. Although Tugwell wanted to concentrate on the RA's long-range objectives, the rehabilitation program soon became the RA's most important work. Farmers who agreed to follow carefully designed home and farm management plans were given small loans, carrying 5 percent interest and secured by crop liens or mortgages on livestock, to purchase equipment, livestock, fertilizer, seed, and other supplies which promised to "rehabilitate" them and to remove their names from relief rolls. In November 1935, when it became apparent that many farmers would not qualify for loans, the RA also began making relief grants, in cash and in kind, to needy farmers. By June 1936, the RA had given grants totaling $15,700,000 and loans totaling $75,600,000 to 635,000 farm families in the United States. In addition to the emergency loans and grants, the RA also took over the supervision of the voluntary state and county debt adjustment committees that had been created during the agricultural crisis in October

Farm machinery buried during a 1935 dust storm. (Kansas State Historical Society)

1933. By the end of 1934 the Farm Credit Administration estimated that 2,600 county committees had been established in forty-four states and that their efforts had saved twenty thousand farms from being foreclosed.

While Rexford Tugwell and the RA worked to improve the plight of poor farmers who lived on the nation's marginal and submarginal lands, Roosevelt created another agency, the Soil Conservation Service, to broaden the Administration's soil-conservation program. New Deal land experts had begun tentative steps to control soil erosion as early as 1933 with the creation of the Civilian Conservation Corps and the Soil Erosion Service in the Department of the Interior. The drought of 1934 brought home to opponents and friends of the New Deal alike the need for a greater emphasis on soil conservation. Government officials estimated that a single dust storm in May 1934 had swept 300 million tons of topsoil from the Great Plains and that water and wind destroyed an estimated 3 billion tons of the nation's soil every year. In 1934 the Soil Erosion Service conducted a survey of the nation's land and found that, exclusive of urban and water areas, only 578 million of nearly 2 billion acres of land suffered no sheet, wind, or gully erosion. Wallace created a Program Planning Division in the Department of Agriculture, under the direction of Howard Tolley, to promote better land-use practices, but the government still lacked a comprehensive soil-conservation program.

The urgent need to make up for lost time was underscored by the dust storms that struck the nation in the spring of 1935. High winds ripped through the West carrying dust from the Plains onto ships sailing far out in the Atlantic Ocean. The worst dust storm in 1935, carried by forty-five- to seventy-mile-an-hour winds, began on April 14 and moved with savage fury across the Southern Plains. When the "black blizzard," carrying topsoil from the Dakotas, western Nebraska, and western and central Kansas, hit Dodge City, Kansas, at 2:40 in the afternoon, the temperature dropped by more than thirty degrees as the storm blotted out the sun and plunged the city into total darkness. As the sky over Washington was blackened by great billowing clouds of dust, the entire nation was finally convinced that something had to be done.

On April 17, 1935, the Soil Conservation Act was passed by the Congress giving the Soil Conservation Service, which was placed under the Department of Agriculture, legislative sanction to conduct surveys and to begin measures to promote soil conservation. The Department of the Interior's soil-erosion programs were also transferred to the Department of Agriculture. Wallace now had the broad powers needed to purchase land, to conduct demonstration projects, to provide aid to individual farmers, and to require that the individual states pass certain laws to prevent future soil erosion.

Wallace had to decide whether to rely upon the extension service and the existing federal bureaucracy, or to create a new administrative structure to govern the program. He had worked through the existing power structure when the AAA was created, but now he decided that to retain control of the SCS it was necessary to bypass the old order by creating new "soil conservation districts" to plan and administer the program. Like the AAA, the SCS was to be decentralized. States were asked to pass legislation creating conservation districts, based upon a model law published by the Department of Agriculture. In many states the American Farm Bureau Federation, the state agricultural colleges, and the extension services, jealous of the SCS's new role, resisted the passage of new legislation. By 1938 twenty-seven states had passed laws to create soil conservation districts; in the Middle West, however, only Michigan, Indiana, and South Dakota had passed laws that were satisfac-

The government planned to resettle thousands of poor farm families on more productive land, but many farmers fiercely resisted efforts to move them from their farmsteads. (Kansas State Historical Society)

tory to the Department of Agriculture. In spite of the obvious need for the program, many states failed to pass the required legislation until the Second World War.

The RA and the SCS concentrated on long-range planning, but the federal government was still forced in 1935 to continue previously developed emergency agricultural relief programs. Drought continued to plague farmers in many sections of the Middle West. In January 1935 C. C. Warburton, director of the Extension Service, reported that farmers in Missouri, Nebraska, Kansas, and Minnesota would again need emergency crop and feed loans. A month later the FCA made $60 million available to meet the continuing credit crisis. During the year in the Middle West 156,276 farmers received emergency crop and feed loans totaling $29,212,766.

In the spring, as farmers began to cultivate their fields, high winds brought one dust storm after another to the Plains. In some areas schools were closed, livestock perished, and people who went outside were forced to wear breathing masks. In Kansas a newspaper correspondent reported that the blowing dust was so thick that ''Lady Godiva could ride thru the streets without even the horse seeing her.'' Some families, beaten by the dust, began to migrate westward in search of a better life. Those who stayed defended life on the Plains and insisted that the rains would soon return. They resented the sensational publicity given to the storms by the Eastern press. Farmers were quick to point out that most of the dust storms did not originate in the Middle West and that only the western edge of the Plains was severely affected by wind erosion. Many scientists agreed. As the nation became more ''dust conscious,'' geologists worked out an easy method to determine the origins of the storms. If the dust was black, it was from northeastern New Mexico; when the dust was golden, from Colorado and western Kansas; and when white, from the alkali beds of northern Nebraska and the Dakotas. Still, as they watched the dust pile up around their homes, none could deny that something drastic needed to be done to stop the flight of the soil from the land.

Fearful that the government might cut back emergency relief programs to discourage farmers from staying on the land, farmers were quick to organize to demand drought relief. In late March the radical Farmers' Emergency Relief Conference sponsored a meeting in Sioux Falls, South Dakota, to discuss the drought situation. Among the five hundred delegates attending were J. J. Schefick, president of the Farmers' Holiday Association in Nebraska, Mrs. Helen Hester of Mitchell County, Kansas, Mrs. Ella Watstad, a member of the United Farmers' League of South Dakota, John G. Walz, secretary of the Wells County Holiday Association of North Dakota, and Lem Harris, secretary of the Farmers' Emergency Relief Conference. Spokesmen at the conference still sounded radical, but their demands were hardly ''revolutionary.''

Radicals at the conference insisted that they had no interest in cost-of-production, mortgage refinancing, or inflation because those issues always became political footballs which ended up helping the rich. The conference did demand, however, the abolition of the AAA, because ''every step on this program has been a history of trading our labor in the interest of banks, insurance companies, speculators and profiteers.'' The conference further demanded that all feed and seed loan obligations be canceled, that government relief allowances be increased, and that farmers be given more credit to repair their homes and to purchase feed, seed, gas, and cattle. To finance the increased cost of agricultural relief the conference suggested the enactment of a steeply graduated income tax on the nation's rich. Pressure was also brought on the railroads to implement

A savage dust storm approaches Alma, Nebraska. (Nebraska State Historical Society)

special rates in the drought areas as they had in 1934. The rhetoric of radicalism remained; but the agrarian left was well on its way to becoming a part of the New Deal consensus.

As the government began to mobilize its resources to combat the drought, May rains turned the "dust bowl" into a rain barrel. Hay and fodder were still being shipped into the drought areas, but the rains saved most of the corn, sorghum, kaffir, milo, and maize crops. With the return of rain, the emergency drought relief programs carried over from 1934 were gradually curtailed. By June 30 only 131 counties in six states, including 36 in Kansas and 4 in Nebraska, were considered emergency drought areas. The drought area expanded somewhat in late summer, particularly in wheat sections of the Western Plains. Roosevelt was optimistic, however, that the FCA, the RA, and the AAA, along with local and state governments, would be able to take care of most of the farmers' needs without additional emergency relief.

Farmers still needing relief who had been previously cared for by the FERA or some other federal agency were now to be cared for by the RA, or the newly created Works Progress Administration. The WPA, under the direction of Harry Hopkins, was created to provide work for the unemployed, but also promised to expand work relief programs in rural areas. Unfortunately, the transition from work relief projects managed by the FERA to the WPA was confused and frequently chaotic. The RA had even greater difficulty creating a viable administrative structure in the field. Although eligible for relief, many families didn't know whom, or where, or when to ask for federal aid. In some cases the old system of relief closed down before the new agencies were ready to go to work. By winter the administrative confusion resulted in considerable suffering in the Middle West. M. W. Torkelson complained to Wallace's Secretary, Paul Appleby, that many farmers in Wisconsin were ''slowly starving to death.'' The Board of County Commissioners in Gray County, Kansas, reported the same conditions to Congressman Clifford Hope. They complained that farmers in their county ''are not able to receive further credit from the stores for groceries; their children are, in various instances, no longer able to attend school because of insufficient clothing to keep them warm going to and from school. . . . Many of the families are on the verge of starvation.'' The RA received most of the blame. Even Harry Hopkins, whose agency was hardly considered a model of administrative efficiency, complained that the RA's organization was so bad that many of its own field agents had not been paid and had to be put on relief. The WPA moved quickly to fill the void. By winter thousands of farmers, who preferred work relief to the dole anyway, were put to work on WPA road and soil-conservation projects.

The AAA also continued to provide immediate cash benefits to farmers. Criticisms continued, but it was also obvious that the AAA now enjoyed considerable support in the farm belt. Senator Arthur Capper reflected the feelings of many farmers and their Congressmen when he wrote to William Allen White, ''The AAA program is an emergency measure and has been a life saver to the bulk of the wheat growers in Kansas. I am not for it as a permanent proposition but it is too soon to throw it down.'' In May nearly four thousand farmers demonstrated in Washington to show their support for the AAA. President Roosevelt, with uncharacteristic bluntness, greeted the crowd and charged his opponents with trying to mislead the American people. The President told the farmers, ''As you know, a great many of the high and mighty — with special axes to grind — have been deliberately trying to mislead the people who know nothing of farming by misrepresenting — no why use a pussyfoot word? — by lying about the kind of farm program under which this nation is operating today.''

The United States still did not have a national land and water policy to manage resources or to protect the public domain when Roosevelt became President. This photograph shows overgrazed land in South Dakota. By the end of the 1930s the Western range was more effectively controlled and regulated by the government. (Nebraska State Historical Society)

Opposition to the AAA was crumbling, but many rural areas were still clearly opposed to the New Deal's overall program. In December 1935 a *Literary Digest* poll showed that nearly two out of three people interviewed in Kansas, Nebraska, Missouri, Iowa, Minnesota, Indiana, Illinois, Michigan, and the Dakotas disapproved of the New Deal. Senator Capper explained that people in the Middle West were especially concerned about the growth of big government, the cost of the federal relief program, and the growing bureaucratic control over agriculture in Washington.

Late in the year, conservatives hoping to defeat Roosevelt in 1936 launched a major propaganda campaign against him in the Middle West. The American Liberty League, which had been organized by a group of conservative businessmen and industrialists to work against Roosevelt's reelection, was particularly vocal in accusing the AAA of regimenting the American farmer. League spokesmen argued that the New Deal's farm program had reduced the farmer to a position "bordering upon peasantry and serfdom." Stanley F. Morse, executive vice president of the Farmers' Independence Council of

America, which had been organized by the Liberty League to mobilize ''grass roots'' sentiment in the farm belt against the New Deal, used the familiar language of the yeoman myth to warn farmers that the Roosevelt Administration was undermining the American character. In a radio address Morse linked the philosophy of the Liberty League to the values of rural America. He appealed to the farmer: ''You and I, whose lives are deeply rooted in the soil of America, no longer can view the moral decline of so many Americans without making an effort to halt this wholesale demoralization. We of the farm where self-reliance, hard work and thrift are essential to success, know that recovery and prosperity never can come from working less, spending more than we earn and leaning on the government.''

Farmers still prided themselves on their individualism, but the American Liberty League made few inroads in the Middle West. Roosevelt remained personally popular and most farmers believed that he had at least tried to deliver on his promise of bringing equality to agriculture. Farmers may not have liked the New Deal, but as newspaper correspondent Roland M. Jones observed, they were also ''inclined to be amused over the fact that people who are worried now about [the farmer's] liberty did not seem to be much worried about his losing his shirt.'' Most important, farmers, particularly in the drought areas, now depended upon the government for survival. Dollars sent their way by the government could still be justified as ''emergency measures.'' Philosophic questions about liberty had little appeal to farmers who happily watched prices rise while waiting for their next check from the government. *Wallace's Farmer* observed, ''Farmers may have lost the liberty to raise ten-cent corn and three-cent hogs, but nobody in the corn belt is going to break down and cry about it.''

Politically, however, the Middle West remained extremely volatile. Farmers did not like production controls and still looked to new markets abroad to find a ''permanent solution'' to the farm crisis. In 1934 Roosevelt and his Secretary of State, Cordell Hull, with strong support from Secretary Wallace, looked for new ways to open the clogged channels of international trade. The result was the Trade Agreements Act of 1934, which allowed the President to raise or lower the tariff by as much as 50 percent by negotiating reciprocal trade agreements. Within the framework of the law, two approaches

to increasing farm exports were possible. Hull favored the uncon-
ditional use of the ''most favored nation clause,'' which envisaged
the multilateral reduction of trade barriers, while Roosevelt's Foreign
Trade Advisor, George N. Peek, wanted to use the act to bargain with
individual countries for bilateral commercial agreements.

In 1934 only one reciprocal treaty — with Cuba — was signed. By
the end of 1935 agreements had been signed with Brazil, Belgium,
Haiti, Sweden, Columbia, Canada, Honduras, and the Netherlands,
but the agreements had not been in effect long enough to have a
measurable impact on farm exports. Peek blamed Hull, rather than
the international situation, for failing to find new markets for American farmers. In November 1935 after a bitter exchange with Roosevelt, Peek finally resigned from the Administration. He quickly
revived proposals supporting McNary-Haugenism and began to
campaign actively in the Middle West against Roosevelt's reelection
in 1936.

In spite of the New Deal's failure to find new markets for American
farm products at home or abroad, the economic position of the
average Middle Western farmer was much improved by the end of
1935. The yearly average for farm prices increased to 108 percent of
the prewar level. The index for grains rose to 103 percent of the
prewar level, dairy products to 108 percent, and meat animals to 118
percent. Allowing for a 17 percent rise in prices that farmers had to
pay for nonagricultural products, farm prices reached 86 percent of
parity by the end of 1935. Cash income again increased in 1935,
totaling nearly $6.9 billion. Cash available for living expenses also
increased slightly, from $3,257 million in 1934 to $3,575 million in
1935.

The government's wheat adjustment program was again influenced by the effects of drought and by a severe outbreak of black
stem rust in the spring wheat areas. Although the government quickly
modified its adjustment contracts to allow farmers to plant up to 165
percent of their base acreage to wheat, wheat production in 1935 was
only 603,100,000 bushels, or 257,371,000 less than the 1928-1932
average. The price of wheat reached 90.1 cents per bushel by late
1935, and cash income from the 1935 crop rose to $485,000,000,
including $115,600,000 in benefit payments. With the wheat carry-
over at the lowest level in years, the economic outlook for wheat
producers appeared to be the brightest since the beginning of the

depression. Officials in the AAA warned, however, that since outlets for the export of wheat remained as restricted as ever, one year of normal weather would again give the United States a burdensome surplus of wheat.

New Deal planners were also concerned by signs that the AAA's voluntary control program was breaking down in many parts of the nation, including the Middle West. Of the sixty-six million acres of land seeded to wheat in 1935, cooperating farmers seeded approximately 41.6 million acres while noncontract signers, who had planted only 14.5 million acres during the base period, planted 24.6 million acres to wheat. In the Middle West wheat farmers had retired nearly four million acres in 1934 but had retired only slightly more than three million acres in 1935.

A corn-hog referendum on October 25, 1935, indicated that more farmers were in favor of the adjustment program than in 1934, but AAA officials were also concerned by the lack of participation in the corn-hog reduction program. Nationally in 1934, 55,310,000 acres had been under contract; in 1935 the number declined to 53,580,000. In the Middle West 10,182,000 acres were withheld from production in 1935, compared to 11,259,000 acres in 1934. Corn production in 1935 was 2,202,852,000 bushels, only 359,295,000 bushels less than the five-year average. Although the average price of hogs had increased to $8.36 per hundredweight, the average price of corn fell to 57.7 cents per bushel. Income from the 1935 crop, including rental and benefit payments, was an estimated $981 million. The government also continued the corn loan program and its surplus removal operations. By December 31, 1935, the AAA had made corn loans, primarily in the Middle West, totaling $6,582,539.25 and had spent $45,623,169 for surplus removal.

The AAA continued to experience difficulties with the dairy industry's marketing agreements. When the National Recovery Administration was declared unconstitutional by the Supreme Court, the AAA's marketing agreements were again rewritten. To stabilize dairy prices the AAA continued to purchase surplus dairy products. From May until the end of December, 7,072,000 pounds of butter, 192,000 pounds of cheese, and 1,316,000 pounds of evaporated milk were purchased by the AAA. The cost of the government's surplus removal operations during the year totaled over $22 million. Efforts to persuade dairy producers to accept a production control program again ended in failure, but dairy prices still increased substantially in 1935. By December 1935 dairy products were selling for 118 percent of the prewar level and at 94 percent of parity.

By the end of 1935 the AAA had still not proved that it could maintain farm prices at near-parity levels without an assist from nature. The RA and the WPA provided valuable emergency relief to farmers, but had not had the time or the resources to mount a major attack against the root causes of rural poverty. Left-wing farm groups were increasingly silent, but conservative forces on the right promised a major campaign against Roosevelt in 1936. Farmers were much better off than when Roosevelt became President, but it was uncertain whether they would vote to continue the New Deal in the upcoming Presidential election.

Chapter 6

Drought and Politics, 1936

President Roosevelt's conservative opponents increasingly looked to the United States Supreme Court to strike down the New Deal's farm program. They scored a major victory when on January 6, 1936, by a vote of six to three, the Court declared the AAA's taxes on processors unconstitutional. The Court argued that the processing tax was not really a tax, but rather was part of a regulatory system designed to control agricultural production. Although the system was voluntary, Justice Owen Roberts, speaking for the majority, reasoned that farmers really had no choice but to accept the payments. Consequently, Roberts concluded, the real purpose of the benefit payments was to ''coerce'' farmers into accepting regulation of the farm economy. The decision was a direct challenge to the entire philosophy of the New Deal. Not only had the processing tax been struck down, the use of government power to control production, and the payment of direct cash benefits to farmers, were now considered unconstitutional.

The invalidation of the AAA met with a storm of disapproval in the Middle West. Morton Taylor, of the *New Republic*, following a two-week trip during which he polled nearly twenty thousand farmers in the Middle West, reported that farmers, even if they had previously opposed Roosevelt and the AAA, were angry with the Court and would support Roosevelt in the upcoming Presidential election. In Iowa farmers displayed their contempt for the decision by hanging in effigy the six justices who had voted against the AAA. The rhetoric of ''rugged individualism'' remained, but it was also clear that farmers, even if they opposed production controls, expected the government to continue some form of agricultural relief.

To respond to the new crisis Wallace called the nation's agricultural leaders to Washington to discuss a new farm relief pro-

A farmer and his son examine wheat damaged by drought on a North Dakota farm, 1936. (State Historical Society of North Dakota)

gram. Wallace now proposed that farmers be paid benefit payments, funded by Congressional appropriations rather than processing taxes, if they agreed to replace soil-depleting crops with soil-conserving crops. Since the major soil-depleting crops in the Middle West were also the major surplus crops, the Administration's plan would not only promote conservation, it would also limit production. It was obvious that the Department of Agriculture had no intention of abandoning its efforts to regulate the farm economy. Wallace compromised on the issue of processing taxes, but farm officials still intended to pay farmers cash benefits and refused to abandon the AAA's philosophy of production controls.

The new Soil Conservation and Domestic Allotment Act won quick approval in the House and the Senate, becoming law on February 29, 1936. The Secretary of Agriculture was also given, under the Supplemental Appropriation Act of February 11, 1936, $296,185,000 to meet the contractual obligations incurred by the AAA before the Court declared the AAA's processing taxes uncon-

Calves were allowed to graze on this dried-up cornfield in North Dakota during the summer of 1936, but many cattle were lost as they wandered off in search of feed and water. (State Historical Society of North Dakota)

stitutional. Although not a single Republican in the House, and only three Republicans in the Senate, voted for the bill, the Congress, like the Department of Agriculture, made it clear that it would not allow the Supreme Court to dictate American farm policy.

Roosevelt's critics worried that the President had again "bought" the farm vote. Still, Roosevelt's opponents on both the left and the right believed that the President could be beaten if they could mobilize the discontented into a new political coalition. In May 1936 the Minnesota Farmer-Labor Party called a national conference to consider establishing a left-wing party to challenge Roosevelt at the polls. When the call for the conference included Communists, a number of important dissident groups, including the Socialists, the Wisconsin Progressive Labor Party, and most of the members of the American Commonwealth Political Federation, refused to attend the conference. Governor Olson, who was dying of cancer, also refused to endorse a third-party Presidential ticket, warning that the result might be the defeat of Roosevelt and the election of a "Fascist Republican." Both the Socialist and the Communist parties nominated candidates in 1936, but neither had a significant impact on the course of the election. The left remained divided into bitter and divisive factions.

Father Charles Coughlin, along with Francis Townsend and Gerald L. K. Smith, who now headed the Share the Wealth Movement, escalated their criticisms of Roosevelt. Their varied appeals still enjoyed widespread support from the old, the poor, the religious, and the discontented on both the left and the right. Although divided on many issues, they were able, primarily through the work of Coughlin, to form a new party, the Union Party, to challenge Roosevelt during the election. Subsequently they persuaded North Dakota Congressman William Lemke to head the Union Party ticket. The party's farm program, which called for inflation, mortgage refinancing, and cost-of-production to end the farm crisis, simply restated the yearly demands of the agrarian left. Lemke, who had a sophisticated understanding of agriculture, realized that he would not be able to win the election but hoped to be able to draw enough votes away from Roosevelt, particularly in the Middle West, to throw the election into the House of Representatives. His political benefactors, however, scared away many of Lemke's potential supporters in the farm belt. Coughlin, Townsend, and Smith all had legitimate criti-

This woman and her child flee the drought in South Dakota in 1936. Most farm families, however, with assistance from the government, were able to survive the droughts and continue their farming operations with the return of fair weather. (Rothstein photo, 1936, South Dakota State Historical Society)

cisms of Roosevelt's farm program, but their abusive, hysterical, and at times antisemitic attacks against the President led to charges that they were no more than native-born fascists intent upon destroying the American way of life. Lemke kept his composure but lost his respect. Even the Farmers' Union and the Farmers' Holiday Association split on whether they should endorse Lemke's candidacy.

The Republicans turned to Alfred M. Landon of Kansas to lead their campaign against Roosevelt. Landon, the only Republican governor west of the Mississippi to survive the New Deal landslide in 1934, occupied a middle ground between the New Deal and old-guard Republicans. He showed surprising strength in the early preelection polls taken in the farm belt. On agricultural issues, Landon generally agreed with Roosevelt's soil-conservation and farm relief program. He also understood the need for continued federal regulation of the farm economy. Consequently, Landon chose not to launch a frontal assault against the New Deal's farm program. Rather, he and his supporters hoped to bring farmers back to the Republican Party by exploiting less vital, but still controversial, issues in the farm belt. The Resettlement Administration's land retirement program, the "radical Brain Trust," charges of waste and extravagance, and the New Deal's reciprocal trade agreements program would be the pivotal "farm" issues debated in 1936 in the Middle West.

One of the most successful criticisms of the New Deal was that it wasted money. A number of New Deal figures, including Secretary Wallace and Harry Hopkins, came under attack for their "radicalism" and "extravagance" in handling relief funds, but Rexford Tugwell, administrator of the Resettlement Administration was singled out for particularly abusive criticisms in the farm belt. Tugwell, popularly dubbed "Rex the Red," continued to push for the retirement of submarginal lands, the resettlement of poor farmers, and the New Deal's community programs. While the left charged that Tugwell was not doing enough for America's rural poor, right-wing opponents of the New Deal charged that Tugwell, the impractical intellectual, was leading farmers down the path of socialism. Lacking a popular base of support in Congress, Tugwell became, as *Wallace's Farmer* observed, the "official goat" for everything conservatives disliked about the New Deal. Roosevelt failed to come to Tugwell's support and forced him to remain on the sidelines during the election campaign. There was simply no political advantage in doing otherwise. Roosevelt was not an ideologue. His interest was in winning the election. To defeat Landon, and to keep the left and right from uniting around a single issue, Roosevelt decided to maintain the middle-of-the-road position he had occupied since the beginning of the depression. Poor farmers were still largely unorganized and could make few political demands on the President.

While conservative farm spokesmen such as Dan Casement, president of the Farmers' Independence Council of America, wanted Landon to focus on the loss of freedom and the "regimentation" of the American farmer, Roosevelt's greatest weakness appeared to be the Administration's reciprocal trade agreements program. Early in the year George Peek, who endorsed Landon, traveled through the Middle West showing samples of fresh Argentine beef, Polish hams, canned corn from Germany, carrots and beans from Italy, and canned

In spite of the drought, the shelterbelt project continued. (Kansas State Historical Society)

sausage from Holland, trying to stir up sentiment in the farm belt against the reciprocal trade agreements program. Peek still opposed production controls and hoped to persuade farmers that the solution to the farm problem was still to be found by developing new farm markets abroad. If he could convince farmers that the Administration had not only failed to find new markets, but also had failed to protect American markets against foreign competition, Peek believed that farmers would turn against Roosevelt. Wallace and Roosevelt defended the Administration's trade program, arguing that it would ultimately create new markets for American farm products. Still they worried that farmers might be persuaded to turn their backs on production controls. Both worried that Roosevelt might even lose the Middle West in the November elections.

Any question about the farm vote was settled when drought again struck the Middle West. Although the drought of 1936 was not nearly as severe as the drought of 1934, the Administration again mobilized an emergency relief program to aid thousands of drought-stricken farmers. Early in 1936 dust storms and high winds again dramatized the need for a continuing program of rural relief. In February Roosevelt made $30 million available to the Farm Credit Administration (FCA) to lend farmers money to seed their crops. When demands for the loans increased in the early months of 1936, an additional $7 million was transferred from the RA to the FCA to finance additional crop production loans. Some 275,000 farmers, however, who were receiving relief from the RA or the WPA, were ruled ineligible for the FCA loans. The ruling infuriated farmers in the Middle West who had found that the best, and frequently the only way, to remove their names from relief rolls was through loans from the FCA. Even farmers who were eligible for the program protested that the arbitrary maximum of $200 which had been established for the loans was insufficient to meet their needs. Their complaint was not, as conservatives charged, that the New Deal was doing too much for farmers, but that it was not doing enough.

On April 1, Daniel W. Bell, director of the budget, warned Roosevelt that the FCA's loan policies were undermining his support in the farm belt. C. C. Talbott, the president of the North Dakota Farmers' Union, also warned that the tangled credit situation was hurting Roosevelt. Talbott astutely observed, "If I know the temper of these farmers, they are not going to feel very kindly disposed

The New Deal failed to find a solution to the "farm problem," but President Roosevelt kept hope alive. Here the President visits with a farmer and his son near Mandan, North Dakota. (State Historical Society of North Dakota)

towards the administration if they are left to go on relief. They will, of course, have plenty of influence on their neighbors, even though they may not have the same needs.'' Talbott continued that although Roosevelt would win the solid South, he needed the vote of the common man in the Middle West to win the election.

Roosevelt moved to contain the mounting protest. On April 28 he wrote to Senator Joseph Robinson that the FCA would modify its ruling and would make loans to farmers who were on the RA's inactive lists. At the same time Roosevelt granted another $2 million to the RA to make emergency crop and feed loans to farmers who were still unable to qualify for the FCA program. By the time the RA program went into effect in early May, however, most of the crops in the Middle West, except for the northern counties of North and South Dakota, had already been planted. As a consequence, only a limited number of farmers in the Middle West benefited from the RA's loan program in 1936.

In May, as rain fell in most of the Middle Western states, the drought situation improved and the important spring feed crop was saved. By the end of June, however, there were again widespread reports of drought damage, particularly in the winter wheat areas of North and South Dakota and in such southern states as Georgia and South Carolina. On June 26 Governor Walter Welford of North Dakota informed Roosevelt that forty of the state's fifty-three counties needed aid. The next day John D. Hazen, the Democratic nominee for governor of North Dakota, also reported that in western North Dakota the drought had destroyed feed crops and pastures and that thousands of farmers were in need of immediate relief. In June and July, as the drought began to spread to other sections of the Middle West, officials in Washington began to express concern that drought, combined with another massive invasion of grasshoppers, might also endanger the corn crop.

On June 22 President Roosevelt began what was to become a mammoth relief campaign when he ordered a survey of drought conditions in the north central and southeastern portions of the United States. A few days later Secretary Wallace appointed a drought aid board to formulate tentative drought relief plans. Officials of the WPA and the RA, the two agencies that would carry the brunt of the relief load, also promised to expand the government's works and direct relief programs in the drought areas.

Roosevelt continued to stress the importance of long-range planning for agriculture. On June 27 he indicated that the government, in addition to providing emergency relief, would also pursue a permanent four-point relief plan in the drought areas. Roosevelt announced that the Administration would continue the activities of the Soil Conservation Service, the Great Plains Shelterbelt Project, the land purchase program of the RA, and the construction of dams and irrigation projects under the supervision of the WPA and the Reclamation Service. To coordinate the government's drought relief program Roosevelt appointed, on June 30, a special drought relief committee, consisting of Wallace, Tugwell, Hopkins, and Daniel Bell.

By late June, as the drought spread to Ohio, Indiana, Illinois, Kansas, and Minnesota, Roosevelt assured farmers that the government had enough funds from the general relief appropriation to assist from fifty to one hundred thousand families in the drought areas. With North Dakota, South Dakota, Minnesota, Montana, and Wyo-

Many farmers supplemented their income in the 1930s by working for federal relief agencies such as the WPA. (State Historical Society of North Dakota)

President Franklin D. Roosevelt talking to Steve Brown, a homesteader, in Jamestown, North Dakota, during his drought area inspection trip in August 1936. (Franklin D. Roosevelt Library)

ming already in need of emergency relief, Hopkins also promised, "We have made no estimates and set no limitations on the amount of money to be allotted."

On July 2 Secretary Wallace, who was touring the Middle West, announced that the government would again purchase cattle in the drought areas. A Federal Livestock Feed Agency, under the supervision of E. O. Pollock of the Agricultural Extension Service, was also established to coordinate information about available feed supplies. The AAA also began to modify its crop control programs to preserve all available feed supplies and to encourage the planting of feed and forage crops. It was soon apparent, however, that the livestock industry would need less assistance than in 1934. A large carryover of hay from the 1935 crop, combined with an abundant spring feed crop in 1936, gave farmers a supply of feed that was greater than the preceding five-year average. Also, as a result of the government's previous reduction programs, there were 14 percent fewer grain-consuming, and 7 percent fewer hay-consuming, animals than there had been during the drought crisis in 1934.

As the scaled-down cattle program got underway, the WPA and the RA also began to develop more detailed drought relief plans. Both agencies emphasized that relief activities would be managed and almost entirely financed from Washington. On July 5 Hopkins announced plans to put drought-stricken farmers to work, primarily digging wells, building earth dams, and constructing farm-to-market roads. Seven thousand farmers in South Dakota, 10,400 in North Dakota, and 2,000 in Minnesota were to be employed by the WPA. The RA, which was already caring for 70,000 farmers in the drought area, also announced plans to provide loans and grants to an additional 130,000.

Hoping to exploit the drought, John D. Hamilton, national chairman of the Republican Party, and members of the Farmers' Holiday Association, spread fears that the AAA's reduction program would cause a national food shortage. Roosevelt, who now began to take personal control of the drought relief effort, assured the nation that there was no food shortage. He was quick to add, however, that if shortages did develop, they would be the result of the drought, not crop reduction, since more crop land had been planted in 1936 than in any year since 1933. Roosevelt also announced that he intended to travel to the Northwest to take a "look see" at the drought areas. The President insisted that the trip would be nonpolitical: "This is too serious a thing to get mixed up in politics."

By the end of the first week in July it was obvious that the drought would have a major impact on the election campaign. By July 10, 16,500 farmers in the drought areas were at work on WPA projects and another 55,000 had been authorized for the works program. Tugwell also announced that the RA would spend $1,698,000 each month for "dole payments" of $20 a month to each suffering farm family and that $9,000,000 would be available for crop loans and $9,600,000 for feed loans. RA officials also decided to declare a one-year moratorium on rural rehabilitation loans which had been given previously to distressed farmers in the drought areas.

In spite of the government's efforts, officials in Washington were flooded with demands for an even greater relief effort. H. A. Mackoff

*Crowds line the street as President Roosevelt and Kansas Governor Alf Landon arrive in Des Moines for the Drought Relief Conference, 1936. (From, the **Des Moines Register**, State Historical Society of Iowa)*

of North Dakota complained that "red tape" was destroying the morale of farmers in the drought areas and suggested that the entire farm population be put to work on government projects. Tugwell promised more aid, but he insisted that the complaints about red tape were unfair and were motivated by politics rather than weaknesses in the program. Secretary Wallace tried to focus attention on the Administration's long-range plans but also promised to speed up the immediate delievery of government allotment checks to the drought areas. Federal officials in Washington were determined not to allow the clamor for emergency relief to postpone their long-range plans for agriculture. On July 22 President Roosevelt created the Great Plains Drought Area Committee. The primary objective of the committee was to provide a comprehensive plan to preserve the agricultural productivity of the Great Plains region. Morris L. Cooke, administrator of the Rural Electrification Administration, was named chairman of the committee.

By late summer, as temperatures soared, New Deal officials worried that the drought might hurt Roosevelt politically unless the Administration accelerated its immediate relief programs. By the end of July 550 counties in sixteen states had been designated for emergency relief. Still, farmers complained that the government was moving too slowly. By early August the WPA had plans to employ over sixty thousand farmers in North Dakota, South Dakota, Minnesota, Kansas, Nebraska, and Missouri; but fewer than thirty thousand, all in North and South Dakota, were actually at work. The RA hurried its program of direct relief, but the Administration's drought relief program had already become a major political issue.

As Roosevelt and Landon stepped up their political campaigns in the latter half of 1936, Republican National Chairman Hamilton publicly charged that the Administration was trying to use drought relief to buy Roosevelt's election. When Wallace defended the Administration by charging that Hamilton knew nothing about agriculture, the latter retorted, ''As far as I am concerned, we are both in the same boat.'' Roosevelt also denied Hamilton's charges, adding, ''It is a very great disservice to Government, as a general proposition, to link up human misery with politics.''

The Roosevelt Administration would have provided drought relief even if it had not been an election year. Since it was an election year, however, they also realized that drought relief could help Roosevelt carry the Middle Western farm vote. R. L. Cochran, governor of Nebraska, initially feared that Roosevelt would lose the Middle West to Landon. Cochran bluntly informed James Farley, ''Our drought

Roosevelt and Landon in Des Moines. The simultaneous arrival was planned to present the presidential candidates as "equals" during the conference. (State Historical Society of Iowa)

conditions in the middle-west, much as we hate to think of them, may be an opportunity, through the granting of assistance to help the situation." Representative Fred Hildebrandt of South Dakota assured Farley that the drought would help the President because farmers realized "that the administration has saved them from near-starvation by its help." Some farmers still believed that the drought was God's way of punishing the New Deal for tampering with the laws of nature, but the vast majority of farmers in the Middle West, particularly in the drought areas, realized that Roosevelt's agricultural relief programs had saved thousands of farmers from complete economic ruin.

It was virtually impossible for Governor Landon to benefit from the drought crisis. As governor, Landon had been on the front lines demanding federal drought relief for Kansas farmers. Many farmers worried, however, whether Landon would continue New Deal relief programs as President. A Democratic campaign pamphlet, "When Drought Comes," pointed out that the Democrats had given relief liberally in the drought areas, but questioned whether the Republicans, given their record under Coolidge and Hoover, could be counted upon to continue to bring relief to the agricultural community. The pamphlet also tried to link Landon with Wall Street, "the same old gang that refused in 1930 to use the federal government to relieve human distress. . . ." The Republicans countered with another pamphlet, "Those Who Need Relief Will Get It," but the Democrats bombarded the Middle West with propaganda that Landon could not be trusted to continue the government's drought relief program.

Roosevelt pressed his advantage in the drought areas even further when he announced on August 7, 1936, that the governors of the drought states, including Governor Landon, would be invited to meet with him in the near future to discuss immediate relief measures, relief for the coming winter and spring, and a program of long-range relief. Landon quickly accepted the invitation and promised to meet with Roosevelt anywhere at any time if it would benefit Kansas farmers. Two weeks later Roosevelt wired the governors that he was coming west to obtain firsthand information concerning the drought areas, and on August 26 he departed for his much-publicized tour of the stricken agricultural regions in the Northwest. Roosevelt still insisted that he was not talking politics on the trip, but all along the way he gave speeches defending the government's works program against charges of waste and extravagance, applauded the government's soil-conservation programs, and assured farmers that they would not be forced to move from their homes to resettlement projects. Most important, the President personally guaranteed that the government would not abandon those farmers who still needed relief.

As the Presidential train moved through the drought areas, Roosevelt stopped for major drought conferences with state and federal officials in Bismarck, North Dakota, Pierre, South Dakota, Springfield, Illinois, and Des Moines, Iowa. Landon was invited to attend the conference in Des Moines, which was scheduled for September 3. The meeting between Landon and Roosevelt was marked by "respectful cordiality." The Kansas governor emphasized the need for state and federal cooperation, commodity loans to conserve feed and seed supplies, work programs, and projects to control soil erosion. For his part Roosevelt raised questions about the severity of the drought and explained existing federal relief programs.

Many of Roosevelt's partisans felt that the drought conference in

President Roosevelt talking with Iowa Governor Clyde Herring and other government representatives from the drought states in Des Moines in 1936. Kansas Governor Alf Landon is in the light-colored suit; Missouri Senator Harry S. Truman is at the extreme right. (State Historical Society of Iowa)

Des Moines sealed Landon's fate in the campaign. Landon handled himself well but ultimately could only promise to continue the relief programs already pioneered by the New Deal. By contrast, Roosevelt's presence in the drought areas served as a constant reminder that the Administration had already acted to fulfill many of the President's promises to the nation's farmers. Drought relief was increasingly viewed by farmers not as charity, but as a logical and legitimate response to their needs. Even if they were living on the "dole," farmers still believed that they had retained their essential dignity and independence. The crisis brought people together and guaranteed that farmers would remain a part of the New Deal coalition, at least until the return of fair weather. Drought relief, and the farmers' dependence upon the government for survival, had become institutionalized. During President Hoover's term in office, the question of whether the federal government should provide drought relief was still a controversial issue. Now it was expected. By the fall of 1936 the WPA was providing work relief to three hundred thousand farmers in the drought areas; the RA had made drought relief grants to an additional three hundred thousand farm families.

Roosevelt gained little information from the trip that was not already available to him before he toured the drought areas. The day after he left for the trip, the Great Plains Drought Area Committee, which had been surveying the situation in the Middle West for several weeks, presented Roosevelt with a comprehensive report of drought conditions and listed the needs of the areas. The President's elaborate political charade also had a negligible impact on existing federal relief programs. When he reported to the nation on his meetings with the drought-states governors, he emphasized that relief would continue along the same lines; again he emphasized that the government would not abandon the farmer.

Roosevelt did, however, create two new committees after his return, a move that would eventually result in the expansion of federal relief activities. On September 29, a few days before Landon advanced a similar proposal, Roosevelt created a committee to recommend a long-term program for conserving the resources of the Great Plains area and to work out a plan for a federally sponsored system of crop insurance. Two days later he created another committee to study the problem of farm tenancy in the United States. Neither would have an impact on federal relief activities in 1936, but by merely appointing the committees, Roosevelt held out the alluring promise of even more government support for agriculture.

After his meeting with Roosevelt, Landon stepped up his campaign to capture the Middle Western farm vote. Landon's dilemma on drought relief also characterized his general campaign pronouncements about the future of American agriculture. Following the Republican line, he criticized the cost of the government's relief program. At the same time, however, he promised not only to continue Roosevelt's programs basically unchanged, but also to expand government relief efforts in rural areas. Many farmers questioned whether Landon, even if he tried, would be able to match Roosevelt's record on agricultural relief. Congressman Charles Brand of Ohio warned farmers that even if Landon wanted to continue Roosevelt's farm program, a vote against the President would be interpreted by Landon's conservative supporters as a repudiation of the New Deal.

Many Middle Western communities had to turn on street lights at midday as the sun was blocked out by the billowing clouds of dust. (Kansas State Historical Society)

Roosevelt, in a series of speeches in the Middle West in October, vigorously defended the Administration's farm program while constantly reminding farmers that they were better off in 1936 than they were in 1932. He proudly took credit for increasing farm income, for saving thousands of farms from foreclosure, and for promoting soil conservation. He insisted that the Reciprocal Trade Agreements Program would pay dividends in the future. Always, the President reminded his audiences of the Administration's exhaustive emergency drought relief program.

The President studiously avoided controversial issues about production control programs and the related question of whether the New Deal's approach to the farm crisis had benefited or harmed the nation's poorest farmers. Had poor farmers been organized, or if Roosevelt had faced a more serious challenge from the left, he could not have avoided such substantive issues. The poor, however, in spite of the RA's efforts, were still an unorganized political force. The challenge from the left continued to disintegrate through the summer and fall. Lemke's candidacy, which had seemed so threatening in the spring, proved to be little more than an annoyance. W. T. Neal, writing to James Farley, observed on the eve of the election that farmers viewed the Union Party campaign much as they would a skyrocket: "It looked beautiful when it first lighted, but they, the thinking ones, commence to realize it will only be a burnt stick in November, and they are filtering back to their old parties. The only hot ones left are the organizers."

Lemke's farm program was popular in the Middle West, but many farmers were afraid that a vote for Lemke would take enough votes away from Roosevelt to allow Landon to win in November. Many listened to Minnesota Governor Olson, who endorsed Roosevelt from his deathbed with the warning, "Liberals must unite in 1936 to reelect Roosevelt to prevent the election of reactionary Alf Landon." Following Olson's lead, the La Follettes in Wisconsin also endorsed Roosevelt, saying, "We must choose between Franklin D. Roosevelt, with his forward looking humanitarian record — and the reactionary Republican candidate chosen and backed by the Du Ponts, Morgan, Hoover, the Liberty League and the Wall Street interests." Roosevelt also won endorsements from other important political leaders in the Middle West, such as Farmer-Laborite Henrik Shipstead of Minnesota and Senator George Norris of Nebraska, who was running as an independent in the Cornhusker State. Only two members of Congress, Representative Usher L. Burdick and Senator Lynn Frazier, both of North Dakota, gave Lemke consistent support during the campaign. Although Lemke had the support of many leaders of the Farmers' Holiday Association and the Farmers' Union, their strength was more than offset by the American Farm Bureau Federation and the National Grange, both of which supported Roosevelt.

In November Roosevelt swept the Middle West, burying Lemke and Landon in one of the worst landslides in history. In the Middle West Roosevelt won 10,349,298 votes to Landon's 6,847,218. Lemke, who won 891,886 votes nationally, failed miserably with farmers. Even in his home state, where he was most successful, Lemke carried only 12.8 percent of the votes cast for the Presidency.

In the wake of Roosevelt's election victory, William A. Hirth, president of the Missouri Farmers' Association, offered an insightful analysis of why Roosevelt had done so well in the farm states. Hirth wrote to Marvin McIntyre, Roosevelt's secretary, "Millions of farm men and women voted for the President not because they are satisfied with the New Deal farm policies, but because they do believe that Roosevelt is deeply friendly to agriculture, and realizing that their

The blowing dust frequently buried the plows and tractors that had plowed under the virgin sod. (Kansas State Historical Society)

general economic position is improving, and remembering that the do nothing policy of Coolidge and Hoover plunged agriculture into the greatest tragedy it has ever known, they were suspicious of Greeks who came bearing gifts. . . .'' Landon could not escape the image that the Republicans had caused the depression and then had done nothing to resolve the crisis. Roosevelt, by contrast, was obviously responsive to agriculture and had emphasized farm relief throughout his first term in office. Conditions were better in 1936 than in 1932. By 1936 the gross farm income, including rental and benefit payments, totaled nearly $9.5 billion, a figure that doubled the income farmers had received in 1932. The average price of farm products in 1936 was 14 percent above the prewar level and had an exchange value of 92 percent of the prewar average. Although the volume of farm exports was still 26 percent below the 1932 total, the value of farm exports in 1935-1936 had again increased, as a result of higher prices, to $766 million.

The improved position of agriculture was also reflected in the activities of the Farm Credit Administration. In October 1933 the FCA had received seventy-seven thousand applications a month for loans to refinance farm mortgages; by 1936 the number of applications had fallen to only seven thousand a month. The rate of mortgage foreclosures, which had averaged thirty-nine per thousand in the spring of 1933, had fallen to half that amount by the spring of 1936. By the end of Roosevelt's first term in office, the government held, as a result of the refinancing efforts of the FCA, nearly 40 percent of the total rural mortgage indebtedness in the United States.

Still, there was widespread suffering in the farm belt, particularly during the winter of 1936. Relief grants from the RA, amounting to only $20 a month, were hardly enough to feed and clothe the rural poor. Many families in the Great Plains, perhaps as many as twenty-five thousand in 1936 alone, were unable to face the hardships of drought and depression any longer and permanently migrated west to Idaho, Washington, Oregon, and California. For many the search for a better life led only to more frustration and despair. Thousands of migrant families, unable to find jobs and lacking money to buy new land, were forced to squat on the outskirts of urban communities in shacks or hovels made of stray boards, tree branches, and strips of tin. By the end of Roosevelt's first term, the problems of the rural poor seemed as enduring as ever.

Nor had the New Deal found a solution to the problem of abundance. The problems facing the agricultural community in 1933 of overproduction, below-parity farm prices, and the loss of foreign markets were obscured in 1936 by the drought and by the government's massive relief program. By 1936, however, the government's acreage control programs, because of nonparticipation and increased acreage yields by participating farmers, were beginning to break down. Drought, although causing untold hardship, rescued the New Deal farm program by limiting surplus production and by keeping farm prices from falling to disastrously low levels. Advances in technology continued to reduce the need for farm labor, driving still more farmers from the land. Roosevelt's massive agricultural relief program from 1933 to 1936 saved many farmers from complete economic ruin and enabled many others, who would have been forced into the ranks of the urban unemployed, to temporarily remain in farming. The long-range causes of the farm crisis in the 1930s would surface again, however, with the return of fair weather. The depression in agriculture would continue to plague farmers and New Deal experts alike until the outbreak of the Second World War.

Drought, depression, grasshoppers, and mechanization drove many farm families from the Middle West in the 1930s. This farm family is hoping to find a new start in Oregon or Washington. (Rothstein photo, 1936, South Dakota State Historical Society)

Chapter 7

The War on Poverty — Fighting a Regiment of Soldiers with a Pop Gun

During the 1936 election campaign Roosevelt promised to do something about the problem of farm tenancy in rural America. The issue was not new; the growth of farm tenancy in the United States had long been a matter of concern. In 1935 a number of southern agrarians, headed by Marvin Jones of Texas and John Bankhead of Alabama, introduced legislation in Congress to address the issue, but, with little support from Roosevelt or the Department of Agriculture, the bill was allowed to die. Resettlement Administration efforts to resettle poor farmers on good land, efforts of the Southern Tenant Farmers' Union to organize tenant farmers in the South, severe drought and dust storms in the Middle West, along with Roosevelt's campaign pronouncements, brought the issue of tenancy to the nation's attention again in 1937. Although tenancy, especially in the Middle West, was not synonymous with poverty, the question of what should be done about tenancy was soon broadened to include a more general discussion of the problems of the nation's rural poor and the Administration's poverty programs.

The depression caused America to look inward and to reexamine its values and institutions. One result was the discovery that rural poverty was widespread in the United States. It did not begin with the depression, but drought and the collapse of the economy exacerbated the problems. In 1935, although the magnitude of the problem was only dimly perceived, the RA began to attack rural poverty, and officials quickly discovered that millions of farm families lived on worn-out land, were heavily in debt, and, because of overcrowding

Leaving the drought-stricken Plains, these South Dakota farmers head for Oregon in 1936. Until the coming of World War II, however, many migrants were unable to find jobs and lived a hand-to-mouth existence. (Rothstein photo, South Dakota State Historical Society)

and primitive farming techniques, had almost no hope for the future. The RA, which spent much of its time providing emergency drought relief during its first year of operation, made little progress in raising the status of the rural poor, but it at least forced the nation to admit that rural poverty was a serious problem in every region of the country.

In November 1936 Roosevelt directed Secretary of Agriculture Wallace to form a citizens committee to examine past and future activities of the RA, to propose solutions for the farm tenancy problem, and to provide a sense of direction for future reform initiatives that might improve the plight of the poor. After several months of intensive investigation and debate, the committee sent its final report to the President. The report noted with alarm that the rate of tenancy, which had stood at only 25 percent in 1880, had increased with each passing decade until by 1935 the percentage of farmers throughout the country who were tenants had risen to nearly 42 percent. In the Middle West the number of tenants had increased in

This Middle Western farm family heads west in a "modern" covered wagon to look for work. (Rothstein photo, 1936, South Dakota State Historical Society)

every decade since the 1890s until by the mid-1930s 821,518 of the region's 2,263,543 farms were operated by tenants. In the Middle Western states tenancy rates ranged as high as 50 percent in Iowa, 49 percent in Nebraska and South Dakota, and 40 percent in Kansas and Illinois.

The committee recognized that the life of the tenant farmer in the Middle West was different politically, economically, and socially from that of his southern counterpart. Poor farmers in the South, especially if they were Black, had no political voice, lacked mobility, and were trapped in a system of racial prejudice and segregation that made fundamental reform almost impossible. Violence, hatred, discrimination, abject poverty, and hopelessness had become a way of life for poor farmers in the South. But their unique situation should not obscure the fact that many farmers in the Middle West were just as poor, facing equally insurmountable problems. Poverty was not as widespread in the Middle West, but it was a serious, longstanding condition.

In the past, farm tenancy had been viewed optimistically as a logical and necessary step up the agricultural ladder of success which would eventually elevate the status of the tenant to that of a private landowner. By the 1930s, with the number of tenants steadily increasing at the rate of about forty thousand a year, tenancy appeared to be a treadmill, rather than a rung on the ladder of success, from which few farmers could reasonably hope to escape. There were a number of immediate causes of the rapid rise in tenancy, including drought, depression, and inadequate farm credit, but the increase was also deeply rooted in two historic patterns that had shaped the development of American agriculture. The first was cyclical periods of boom and bust, triggered by overproduction, which caused marginal farmers to lose their land and fall to the rank of tenant farmer or, increasingly, into the rank of the migratory-casual worker. The second closely related trend, which gained momentum as the twentieth century progressed, was the modernization of agriculture. As agriculture became more complex and more highly mechanized, the capital investment needed to be a successful farmer increased dramatically. Agriculture was not yet controlled by a landed aristocracy, but the Jeffersonian promise of an independent, self-reliant, yeoman class was dying.

The Committee on Farm Tenancy recognized that the United States was at the end of an era, but its proposed solutions to the rise in tenancy, and to the problems facing the rural poor, did not deviate from previously established patterns of thought about American agriculture. As Roosevelt had hoped, the committee endorsed the RA's programs for rural rehabilitation, for the improvement of the education and health care of the rural poor, for more extensive plans to retire submarginal land from production, and for the resettlement of marginal and displaced farmers on more productive land. The committee's most important, but fruitless, recommendation was that the government purchase farmland to resell to tenant farmers who wished to become individual landowners. The committee, understandably, linked two problems — land abuse and rural poverty — and tried to solve both at the same time. Consequently its members reasoned that poverty could be ended only if a harmonious balance between man and nature could be restored to agriculture. The committee attacked absentee landlords and concluded that until the tenant farmer (who in many cases moved every year) owned his farm, he

Drought refugee from South Dakota on a highway in Montana making his way west. (Rothstein photo, 1936, South Dakota State Historical Society)

would have little interest in soil conservation and few incentives to make general and long-range improvements in his farming operations.

The committee's recommendations were unrealistic and were doomed to failure from the beginning. The proposed land policy could work only in a frontier society with an open frontier, not in an industrialized society where such a large percentage of the farmers were already tenants. Only the expenditure of billions of dollars, which was economically unfeasible, or the radical redistribution of the land along more democratic lines, which was politically impossible, could have made the nation's tenant farmers individual landowners and restored the independence so desired by the committee. The other recommendations, which would have increased the farmers' efficiency, would have only compounded the already seemingly unresolvable problem of overproduction.

Farm workers replaced by machinery roamed the land looking for work in the 1930s. The government constructed a number of migrant labor camps to partially alleviate their suffering. Pictured here is a transient camp south of Chadron, Nebraska. (Nebraska State Historical Society)

Other alternatives were explored but, given the context of the times, could not be seriously debated. The nation, unless it was willing to turn its back on technology, could simply have conceded that there were too many farmers and that the best solution was to retrain them for jobs in the city. To do so, however, would have meant the abandonment of the mythos about the importance of agriculture to America's future. Culturally, few were willing to give up the idea that a large yeoman class was vital to the survival of American democracy. Also, with unemployment rates still high, there were no jobs in the city for transplanted farmers. Another possibility was to abandon the idea of private land ownership entirely and to encourage the development of collective farms. The New Deal did establish a number of experimental "cooperative farms," especially in the South, but they proved to be unpopular with farmers, who wanted to own their own land, and with the Congress, which feared that the collectivization of agriculture might lead the United States down the path followed by Stalin and the Soviet Union in the 1930s. Since the collectivization of Soviet agriculture had resulted in the deaths and displacement of millions of Russian peasants, the Communist model attracted the support of only a few dreamers. Finally, on a purely pragmatic level, the committee might have accepted the view that tenant farmers were members of a new class in agriculture who were unlikely ever to own their own land. The government, by regulating the contracts between landlord and tenants, could have used its power to protect the tenants' civil rights, to insist on certain conservation standards, and — most important — to guarantee poor farmers a decent standard of living. Few were willing to accept the new reality. To do so would have admitted that the capitalist model of private land ownership had failed. Although the question of what should be done, or indeed in the minds of some Middle Western Congressmen like Clifford Hope whether anything of substance *could* be done, had not been resolved by the committee, Roosevelt quickly took the popular political position that tenant farming was "un-American."

In Congress the call for legislation to deal with farm tenancy came from Congressmen from the South, not the Middle West. The battle for new legislation was again directed, as it had been in 1935, by Marvin Jones in the House and by John Bankhead in the Senate. Middle Western Congressmen appeared reluctant to begin any new programs that would require more federal spending and seemed to prefer the view that the problem of rural poverty was a uniquely "southern" problem.

Middle Western farm spokesmen worried less about poor tenant farmers, who remained relatively quiet, than about private landowners who were still struggling to get back on their feet. Henry C. Taylor of the Farm Foundation, Earl C. Smith, the vice president of the Kansas Farm Bureau Federation, and O. O. Wolf, president of the Kansas Farm Bureau, argued that the best way to help tenant farmers was to raise farm prices by improving the general agricultural situation. They worried that any new legislation that tried to help poor farmers would only increase competition and would eventually undermine the Roosevelt Administration's efforts to raise prices by controlling surplus agricultural production. Other farm spokesmen, such as Kansas Senator Arthur Capper, worried that "undeserving" tenants might benefit from any new federally subsidized program of relief; while his colleague Clifford Hope, who believed that larger farming units were the inevitable wave of the future, warned that the

Some surplus food was distributed to the poor and needy. Here a farmer picks up supplies distributed in Columbus, Nebraska. (Nebraska State Historical Society)

nation's most inefficient farmers would be the primary beneficiaries of the tenant land purchase program. Still another concern was that "class" legislation might encourage social revolution by challenging the existing hierarchy of power and responsibility in the farm belt. Edward O'Neal, who had served on the Committee on Farm Tenancy, had supported the Administration's domestic allotment program, but he now feared that new programs increasing the role of the government in agriculture would undermine the power of the American Farm Bureau Federation, the extension services, and the land-grant colleges.

If Middle Western Congressmen were reluctant to launch a war on poverty, their self-professed agrarian ideologies made it impossible to openly resist efforts that promised to increase the number of free and independent landowners in agriculture. Consequently the Bankhead-Jones Farm Tenant Act moved quickly through Congress with only limited debate. The bill passed in the Senate without a recorded vote and in the House by a vote of 307 to 25. Only four Representatives from the Middle West joined with the minority to vote against the bill. On July 22, 1937, Roosevelt signed the bill into law.

The new law authorized government loans, bearing 3 percent

interest amortized over a period of forty years, totaling $10 million the first year, $25 million the second, and $50 million for each following year. Applications for the loans were to be passed upon by local three-man committees appointed by the Secretary of Agriculture. To satisfy the Administration's conservative critics, the Bankhead-Jones Act provided that tenant farmers, rather than the government, would decide what land should be purchased to avoid ''forced'' resettlement. Also the law effectively excluded the nation's poorest farmers by stipulating that preference would be granted to those tenants who could afford to make a down payment, owned their own livestock, and were generally reliable members of their communities. To prevent land speculation and to guarantee that the government would play a major role in supervising the tenants' farming operations, the law further provided that tenants who received loans from the government would not be allowed to sell their farms for a period of at least five years.

From the beginning it was obvious that, unless the government was willing to spend a good deal more in the future, the loan program would not even be able to keep the number of tenants in the country even, let alone decrease their number. The land purchase program was, without adequate funding, as Clifford Hope suggested, ''like trying to fight a regiment of soldiers with a pop gun.'' William Lemke, who along with Gerald Boileau of Wisconsin had been

Many farmers who left agriculture in the 1930s were driven out by drought, grasshoppers, and low farm prices. Many others, were "tractored out" by labor saving machinery that reduced the number of people needed to feed the nation. (Nebraska State Historical Society)

among the few Congressmen from the Middle West to demand that the government spend as much as $500 million a year to finance the program, blasted the Congress for its failure to come to grips with the magnitude of the problem. Lemke admonished Congress: "I am surprised to hear so much fuss about nothing. If ever a mountain labored and produced a mouse, this bill is it. We have heard a lot of lip service that we are going to make farm tenants farm owners. In the light of that lip service, this bill is a joke and a camouflage."

In the long run the most important provisions of the Bankhead-Jones Tenant Act were those which called, as Roosevelt had wanted, for the continuation of the Administration's general "war" on rural poverty. The law provided for the extension of the Resettlement Administration's emergency relief activities and for the continuation of its more permanent rehabilitation programs. The bill also provided $10 million for the fiscal year ending June 30, 1938, and $20 million in each of the next two years, to continue the Administration's controversial land retirement and relocation program. To coordinate the land retirement plans, Wallace transferred the responsibility for land purchases to the Bureau of Agricultural Economics in the Department of Agriculture. He also changed the name of the politically unpopular Resettlement Administration to the Farm Security Administration (FSA). Rexford Tugwell, who resigned from government service after the 1936 election, was succeeded as head of the new agency by Will Alexander.

While Roosevelt's left-wing opponents pointed out that the New Deal's poverty programs were largely ceremonial, a comparative study of the attitudes of corn and cotton belt farmers by E. A. Schuler, which was released by the government in 1938, suggests that the Administration's approach to the problem of rural poverty and insecure land tenure had widespread support, at least among poor farmers in the corn belt. Landless farmers expressed little bitterness about their position because they believed they would become landowners in the near future. With the exception of the AAA, which was unpopular with virtually every class in American agriculture, New Deal programs were popular among tenants, croppers, and farm laborers in the Middle West. In spite of government statistics to the contrary, Schuler's study indicated that the disadvantaged classes in the corn belt believed that their economic position was improving, rather than declining. The survey revealed that only 3 percent of the

Soil blown onto the roof and side of a building in North Dakota during the 1930s. (State Historical Society of North Dakota)

full owners, 1 percent of the part owners, and 26 percent of the laborers felt that they were worse off than any other class of Americans during the depression.

Farmers still wanted to own their own land, but less than one-half believed, given the realities of the farm crisis, that they would be appreciably better off as landowners. Conflicts between landlords and tenants were probably obscured by Schuler's study, but it is apparent that there was a relative absence of agrarian class consciousness in the Middle West by the late 1930s. Farmers in the Middle West shared a common faith in the virtues of private landownership. Most also shared similar economic positions. Schuler's study indicated that the median income for farmers who owned their land was $1,639, for renters who were related to the owner $1,323, for renters unrelated to the landlord $1,462, for related sharecroppers $1,500 and for unrelated croppers $937. Farm laborers, who earned an average of only $400 to $450 a year, were the worst off, but they too still believed they could become independent landowners. Few farmers understood that modern technology and the rise of large-scale farming techniques would radically alter traditional land-holding patterns in the Middle West and would ultimately undermine the agrarian civilization that the Roosevelt Administration had pledged itself to protect and defend.

Middle Western farm organizations made it clear that programs to help the rural poor should not result in cutting support for other farmers. In 1937 Roosevelt made a number of moves to try to balance the federal budget. The President pointed out that cheap interest rates for agriculture cost the government $40 million a year. When he tried to raise existing interest rates, which ranged from 3 to 3.5 percent, to 4 percent, there was a storm of protest in the Middle West. Farm spokesmen argued that higher interest rates would bankrupt many farmers and would increase the spread of farm tenancy. The farm bloc had enough support to defeat Roosevelt's efforts in 1937, and in 1938 and 1939 when Roosevelt again tried to raise interest rates. On the eve of the Second World War farmers who owned their own land were receiving credit from the government on terms that were almost as liberal as those provided for landless farmers under the Bankhead-Jones Farm Tenant Act.

Mortgage foreclosures were still a serious problem. Farmers scored another major victory when the Frazier-Lemke farm mortgage bill again passed Congress. The bill provided that farmers who were about to lose their farms could declare bankruptcy and prevent foreclosure proceedings for a period of three years. Although the bill, unlike previous versions, was not mandatory and could be put into effect only after a judge had ruled on the merits of each individual case, it represented, nonetheless, an important advance for the farm community.

At the same time that the government developed programs which promised to keep farmers on the land, the Roosevelt Administration also argued that there were too many farmers trying to make a living in the Great Plains region. On February 10, 1937, Roosevelt forwarded to Congress a report of the Great Plains Committee, which he had appointed in September of the previous year, with an endorsement of the committee's recommendations. The report, *The Future of the Great Plains*, estimated that about 40,000 families, or about 165,000 people, had migrated from the Plains states since drought

Earth dam. (Nebraska State Historical Society)

began to strike the area in 1930. Alarmingly, the report predicted that as many as another 500,000 people might have to be "resettled" before a proper balance between man and nature could be restored in the area. The committee suggested that sound rangeland techniques, improved methods of conservation, and the development of water resources would help stabilize the farm economy in the Great Plains, but still concluded that the government should purchase, and retire from production, at least twenty-four million acres of submarginal land in the Plains region. For those farmers who remained on the land, the committee emphasized that they needed to increase the average size of their landholdings and become more efficient if they hoped to continue farming in the future. To help farmers increase the size of their holdings, the committee further recommended that the government purchase, and then resell on liberal terms, another seven million acres of productive farmland in the Great Plains region. By January 31, 1938, the government had purchased 9,127,403 acres of land, 5,149,812 of which was in the Great Plains.

While the government planned for the future, thousands of farmers continued to flee the Plains region. A year after the committee issued its report, M. L. Wilson estimated that 58,400 families, with an average of five members in each family, had left the Great Plains, and that another 40,000 would soon follow because they could not be self-supporting on the land. Driven from the Plains by the combined forces of drought, depression, and modernization, migrant families faced a bleak future. Wilson estimated that, of the more than 58,000 families who had migrated, less than 3,000, or about 5 percent, had managed to relocate where they could be self-supporting. With the number of unemployed hovering between eight and nine million for the country as a whole, few displaced farmers found jobs in industry. Some farmers worked as farm laborers, but their jobs were usually seasonal and offered little security. Lawrence Westbrook, worried about the plight of dispossessed farmers in the South and the West, wrote to Harry Hopkins that thousands of farmers were without homes or jobs and, wandering from place to place, were exploited by employers, ignored by public officials, and were rapidly "degrading into the untouchables of our social and economic life."

Secretary Wallace admitted that the Administration had been "slow" in recognizing the problems of the tenant and the farm laborer. He warned Roosevelt, "This whole problem has in it so

A number of federal agencies, but especially the RA, tried to retire millions of acres of submarginal land from production in the 1930s. This scene near Granville, North Dakota, shows why many farm experts worried that the Middle West was becoming a "Great American desert." (State Historical Society of North Dakota)

much dynamite, political and otherwise, that we cannot neglect it much longer." There was, however, no consensus about what should be done. The newly created FSA was again caught in the middle. John Fisher, the director of information for the FSA, captured the frustration of many officials in the FSA when he wrote to C. B. Baldwin, the assistant administrator, "Claims are made that the loans are too small; that the loans are too large, and that we could help twice as many if we loaned half as much per individual; that we are not demanding sufficient guarantees; that we are demanding too large a share of the borrowers' income; that we are attempting to make farmers from a group that never farmed successfully; that we are regimenting the borrowers when we provide supervision."

The FSA, like the RA before it, was also forced to divert much of its attention away from its long-range objectives to once again provide emergency relief to the drought areas. Many farmers, in spite of the continuing drought, clung tenaciously to their farms. Most of the Middle West enjoyed favorable weather in 1937 and 1938, but drought still plagued a number of farming areas. On July 6 the presidents of the state Farmers' Unions in Kansas, North and South Dakota, Nebraska, and Montana met to demand that the Administration give loans, cash grants, and work relief for farmers who faced still another year of hardship in the drought-stricken regions. John Vesecky, who had been elected as the national president of the Farmers' Union, assured Roosevelt, "None of the folks who took part in the conference which brought out these recommendations can be accused of being calamity howlers or false alarmists. They have been, more or less, consistent supporters of the progressive acts of your administration."

As reports of suffering increased, the FSA, the WPA, and the FCA again made plans to modify their programs. Administration officials, anxious to implement long-range programs, were obviously frustrated by the perpetual need for emergency relief in the drought areas. Undersecretary of Agriculture M. L. Wilson made it clear that many of the emergency programs, which allowed thousands of marginal farmers to stay on the land, were inconsistent with the Administration's long-range plans to retire land from production and to resettle large numbers of Plains farmers. Wilson also realized that it was politically impossible to reverse the government policy of providing drought relief to alleviate human suffering. The best the

This North Dakota farmstead shows the devastating consequences of soil erosion in the 1930s. (State Historical Society of North Dakota)

government could do was to use the relief program to encourage conservation practices, to develop water resources, and to promote sound land management practices in the drought areas.

In spite of the drought, political controversy, and conflicting goals, the FSA made a number of promising advances in 1937 and 1938. By the late 1930s, 1,700,000 farm families, representing nearly eight million people, still earned less than $500 a year. The FSA continued to assume that one of the most effective ways to attack rural poverty was through a liberal credit policy. In the first year and a half of its existence the FSA made standard rehabilitation loans, totaling more than $118 million, and emergency loans to more than 400,000 families in the drought areas. The FSA also continued to work to scale down the level of indebtedness among the nation's poverty-stricken farmers by continuing the activities of the debt adjustment committees. In 1937 and 1938 the FSA's debt adjustment committees handled 43,674 individual cases, involving an indebtedness of $152,782,640, and managed to scale down the debts by $39,046,328. For various types of agricultural organizations, many of which faced bankruptcy, the FSA also successfully reduced debts totaling $7,929,925 to $3,059,344.

The much publicized tenant purchase program also began. By the end of 1938 the FSA had received nearly thirty-eight thousand loan applications. Nationally by the end of 1938 the FSA, with a budget of only $10 million, had made only 1,887 loans, totaling $9,225,083. The FSA also tried to help farmers who remained tenants to negotiate more secure leases from their landlords. By 1938, 139,000 tenants had improved their tenure status, primarily by obtaining written leases or, since nearly a third of the tenants were forced to move every year, by guaranteeing that they would be able to stay on the same farm for more than one year. For migratory-casual workers the FSA also began to develop migrant labor camps. The camps were supposed to give farm laborers decent housing as they moved from one job to another; but most of the nation's migrants continued to live from hand to mouth as they roamed the country in search of work.

The FSA's efforts to help tenant farmers received a major assist from the AAA. In 1938 the AAA ruled that farmers who reduced the number of tenants on their farms to receive larger benefit payments from the AAA would not receive benefit payments higher than those received under the 1937 program. The ruling proved difficult to enforce, since it was almost impossible to prove that landlords had forced their tenants off the land because of the AAA's crop control program. Still, many tenants were encouraged that the AAA had at least demonstrated some awareness of the problems faced by poor tenant farmers.

To guarantee that the government's money would not be "squandered," the FSA's field representatives closely supervised their clients and tried to teach them new farming methods that would increase their productivity and make them less reliant on the outside world for survival. The FSA encouraged farm families to grow larger home gardens, to can more food, to manage their money efficiently, and to rely less on a single cash crop to insure that they would become self-supporting and would not be forced to depend upon the government's relief programs to make it through every year. Of particular importance to many farmers were the efforts of the FSA to build a network of small rural cooperatives that would, it was hoped, allow poor farmers to have more effective bargaining power when they sold their products and to receive discounted rates when they purchased supplies, tools, and machinery for their farms. By 1939, sixteen thousand cooperatives had been created.

The FSA also worked to improve the health care available to poor

Even in times of abundant harvests, grasshoppers plagued farmers, as this 1938 photograph of a North Dakota farm shows. (State Historical Society of North Dakota)

farmers. The FSA realized that many of the poor, who were often dismissed as being lazy and shiftless, were merely sick and hungry. To restore their health, the FSA tried to make medical care available and affordable, working out a program with state and local medical societies that would provide medical care for thousands of destitute farmers. Under the program, farmers joined medical cooperatives and were required to pay a maximum fee for one year, usually about twenty-four dollars, which was deposited in a general fund to be administered by a trustee approved by the FSA. Doctors, after they treated patients covered by the program, submitted their bills to the trustee for payment. If the bills for the year totaled more than the total amount paid in by the farmers who participated in the program, the money that had been paid was distributed proportionately to the doctors to guarantee that all would receive at least some payment. Although the amount of money received by doctors under the program would total only 60 percent of the bills they submitted for payment, most doctors still had a higher collection rate than they would normally have, and the medical care for poor farm families was much improved.

For the long term, the FSA tried to teach farmers sound land-management techniques to conserve and protect the soil. By the mid-1930s the government estimated that fifty million acres of farmland had already been ruined and that another three hundred million acres had been severely damaged, an area that when combined was larger than the states of Iowa, Illinois, Ohio, Indiana, Wisconsin, and Missouri. In 1937 the FSA released a movie, ''The Plow that Broke the Plains,'' to graphically illustrate the tragic consequences of exploiting the land. The film emphasized that much of the land put into production during the great ''plow up'' during and after World War I should never have been used for farming. The great dust storms of the 1930s, the film suggested, were man-made. The government's previous land policy was faulted, but farmers seemed to receive most of the blame for expanding production into marginal areas and for carelessly neglecting the land. The movie, which was seen by more than ten million people in 1937, was banned in a number of Middle Western communities for presenting a ''distorted'' picture of life on the Plains. Many farmers argued that drought, not unsound farming techniques, had caused the dust storms. They resented the charge that they had not taken care of the land and were quick to point out that much of the blowing dust came from federally owned land that had been neglected by the government. For farmers who practiced sound conservation techniques, the film was seen as a sinister plot by the government to justify its resettlement program and to convince the nation that there were too many farmers, particularly in the West.

By January 15, 1937, farm prices were 131 percent of parity, the highest they had been since June 1930. Secretary Wallace, and most farm spokesmen in the Middle West, however, worried that the Soil Conservation and Domestic Allotment Act, which had been intended to be a temporary measure, needed to be replaced by permanent legislation. Drought had focused attention on the problems of tenants and the rural poor. The return of fair weather would again force the nation to confront the dread of plenty.

Chapter 8

An Ever-Normal, or Abnormal, Granary

In February 1937 Secretary of Agriculture Henry Wallace announced that he would call fifty of the country's most important farm leaders to Washington to discuss and formulate new farm legislation. Unlike 1933, when a pliant Congress had accepted the first Agricultural Adjustment Act with limited debate, or 1936, when Congress voted for the Soil Conservation and Domestic Allotment Act less than two months after the Supreme Court struck down the AAA, debate over the second agricultural act raged for more than a year. The eventual passage of the Agricultural Adjustment Act of 1938 was an important watershed in the nation's history. The philosophy and programs embodied in the legislation would structure American farm policy for the next half-century. The passage of the farm bill was one of Roosevelt's most significant victories, but it also opened wounds and ultimately signaled the end of Roosevelt's political dominance in the Middle West. Debate on the bill was one of the most bitter and divisive of the entire decade.

Wallace set the framework for debate on the farm bill when he indicated that the primary goal of the new legislation would be to establish an "ever-normal granary." He explained that an ever-normal granary would allow crop surpluses, rather than being dumped on the market, to be carried over in years of abundant harvests until they were needed in lean years. To implement the plan, the Secretary proposed that in years when there were huge crops and low prices, such as during the early 1920s, the Commodity Credit Corporation (CCC) would grant loans to farmers to enable them to store their products in the nation's granaries. Then, in years when crops were small and prices were high, like the drought years in the

Henry A. Wallace, Roosevelt's Secretary of Agriculture during the 1930s, address-ing a crowd of Middle Western farmers. (State Historical Society of Iowa)

1930s, farmers could repay their loans and sell their crops for a reasonable profit. Wallace, who drew heavily on the historical precedents of the Confucians in China, the Mormons in Utah, the Biblical Joseph in Egypt, and the experience of Hoover's Federal Farm Board, also made it clear that an ever-normal granary would not end the Administration's soil-conservation programs, nor would it end the AAA's efforts to control agricultural production. Indeed, he argued, the ever-normal granary would work only if the Admin-istration's system of crop controls was continued in the future. Wallace, along with planners in the Department of Agriculture, still had faith that scientific planning would create a society based upon "progressive balanced abundance."

A number of diverse factors made quick passage of a new farm bill impossible. Without an absolute emergency, farmers lacked a uni-fying issue to bring them together. Middle Western Congressmen again revived old cost-of-production plans and the McNary-Haugen proposals, dividing farmers, the Congress, and the various farm organizations into hostile camps. Concerns that the farm bill would cost too much money, that the AAA primarily benefited wealthy farmers, that agricultural policy was determined not by "dirt farm-

Overproduction plagued farmers throughout the 1930s. Here an abundant corn crop is stored at Casselton, North Dakota. (State Historical Society of North Dakota)

ers'' but by bureaucrats in Washington, and that farmers never really favored production controls, were voiced throughout the Middle West. The New Deal coalition also began to break down. During Roosevelt's second term the President was forced to pay more attention to the needs of urban and industrial America. The result was a fierce competition between rural and urban areas for federal dollars. Since the primary purpose of the farm bill was to keep farm prices high, urban spokesmen began to publicly question why farmers should be guaranteed high prices for ''competitive'' products when the end result was increased food costs for people in the city. For the first time since the depression began, the farm bloc's power appeared to be disintegrating.

The primary issue debated in 1937 and 1938 was once again production controls. The 1936 drought had eliminated surpluses, and farm prices in early 1937 were generally high. Gross farm income in 1937 totaled more than $10 billion, and real farm income more than $8.6 billion. Farm prices, with the exception of corn, which averaged only slightly more than 51 cents a bushel during the year, were at near-parity levels and guaranteed most farmers a higher standard of living than in any year since 1929. Forecasts indicated, however, that bumper crops in 1937 would cause the bottom to fall out of farm markets.

Edward O'Neal, of the American Farm Bureau Federation, and M. W. Thatcher, of the Farmers' Union, pushed for a farm bill with mandatory production controls, to respond to the impending crisis. They argued that under certain conditions marketing quotas, which would be enforced by heavy penalty taxes on the amount farmers sold above the limits outlined by the government, should be imposed by the Secretary of Agriculture. Marketing quotas were not new; they had been used to control the production of a number of special crops in the past; but the proposed bill would have made mandatory production controls a permanent part of national farm policy and would have extended marketing quotas to a number of major crops for the first time. More important, the acceptance of mandatory controls was seen by many New Deal opponents as the final step in the total regimentation of the farmer. A number of farm organizations, including the Grange, immediately opposed more controls. Along with other conservative groups the Grange argued that coercion, collectivization, and the end of the Constitution would follow if mandatory controls were included in the farm bill.

The legislative impasse ended when farmers were again forced to admit the reality of unwanted abundance. Without direct crop controls, agricultural production skyrocketed in 1937. Many farmers, convinced that the drought had ended the need for production controls, refused to participate in the AAA's soil-conservation program. Only about 263 million acres of farm land, or about 65 percent of the national total, were covered by the AAA program during the year. The 3,020,037 farmers who did participate in the AAA program received $367 million in conservation payments, nearly 70 percent of which was paid to farmers for diverting "soil depleting" crops from production. In the Middle West, agricultural conservation payments from January 2, 1937, to June 30, 1938, totaled $120,158,725 and rental and benefit payments, which had been contracted before the first AAA had been declared unconstitutional, $7,776,641.

Without direct production controls, and with good weather, farmers in 1937 again produced gigantic surpluses. By late summer, government officials released figures indicating that the corn crop would be the largest since 1919 and the wheat crop the largest since 1931. More than 96,500,000 acres of corn were planted, which was

Few Middle Western farms were irrigated in the 1930s, a fact which made agriculture especially vulnerable to drought. Here a farmer, looking to the future, completes a well irrigation system. (State Historical Society of North Dakota)

actually less than in 1936, but the corn crop totaled 2,651,000 bushels, 77 percent more than in 1936 and 350,000,000 bushels more than was normally needed to meet the domestic and foreign demand. The acreage planted to wheat totaled 81,363,000, 12 million acres more than in 1936, and produced a crop totaling 873,914,000 bushels. There was an immediate demand that Roosevelt make liberal loans to farmers through the Commodity Credit Corporation to put a floor under farm prices. Roosevelt refused to provide commodity loans, however, unless Congress in turn agreed to support the Administration's plan for a new system of crop controls.

Gradually, although heated debate continued, farm spokesmen, first from the cotton states in the South, and then from the Middle West, fell into line. On February 16, two days after he received the farm bill from Congress, Roosevelt signed the Agricultural Adjustment Act of 1938 into law. To implement the ever-normal granary, the new law provided that in years when there were huge crops the CCC, which had been used on a limited basis for the same purpose since 1933, would grant loans to farmers to enable them to withhold their products from market. Then, in years when crops were small and prices were high, farmers could repay their loans and sell their crops for a comfortable profit. The loans were to be given in years when the price of certain crops, such as wheat, cotton and corn, fell below a certain predetermined percentage of the parity price. If, however, farm prices did not rise, the government absorbed the loss and retained possession of the crops that the farmers had used for collateral.

If the granary were to be kept from bursting, it was essential that the government continue production controls. Consequently, the new law gave the government the power to set acreage allotments each year for those crops that were covered by the CCC's loan program. Farmers who agreed to limit their acreage would receive "conservation" payments, as they had since 1936, from the government. If farmers exceeded the acreage limits established by the government, the 1938 law provided that under certain circumstances, the AAA could establish "marketing quotas" to limit the amount of farm products farmers could sell on the open market. The quotas, however, could not be put into effect without the approval, in a referendum vote, of at least two-thirds of the farmers concerned. If the quotas were approved, any farmer who sold more than his quota

Farmers with irrigated land often enjoyed high prices and abundant harvests, in spite of drought and depression, as this irrigated alfalfa field in McKenzie County, North Dakota, demonstrates. (State Historical Society of North Dakota)

would face a heavy penalty tax on the amount he sold above the limits outlined by the government. If, however, farmers chose to sell as much as they wanted on the open market and rejected the Secretary's request for marketing quotas, the bill provided that no loans on the crops involved in the referendum could be made by the CCC. Except in years when farmers rejected marketing quotas, loans from the CCC to the farmers were mandatory, at from 52 to 75 percent of parity. The bill also provided that if farm prices continued to be low in spite of the new program, the Congress could also appropriate additional money to give farmers "parity payments" to supplement farm income.

Many other provisions of the bill, which were really separate pieces of legislation, generated little controversy but nonetheless represented important advances for the farm community. The bill provided for the continuation of the Administration's soil-conservation programs, the extension of the surplus removal operations of the Federal Surplus Commodities Corporation, the creation of regional laboratories to search for new uses of farm products, and a program of federal crop insurance for wheat producers. The bill created the Federal Crop Insurance Corporation, with a capitalization of $100 million, and offered federal crop insurance, on an experimental basis, for the first time for the 1939 wheat crop. Farmers who wished to participate in the program could pay their premiums in cash or wheat, or out of future soil-conservation payments, and could insure their crops for from one-half to three-fourths of their normal acreage yields. In 1939, 165,000 wheat farmers bought government insurance, 55,000 of whom ultimately received indemnities, paid in wheat, from the government. Since the government suffered heavy losses, tentative plans to extend the crop insurance program to other crops were postponed until the Administration had time to determine whether the plan was financially feasible.

To administer the new farm program Wallace moved to reorganize the Department of Agriculture. The government's decision in 1933 to rely on the extension services to administer the AAA's programs led to the frequent charge that the Department of Agriculture was dominated by the American Farm Bureau Federation. The AFBF and the land-grant colleges were reluctant, however, to give up their power. After several years of bitter fighting, a compromise agreement was finally reached in July 1938. Under the agreement it was decided that the Department of Agriculture would have its own independent network of state and county AAA offices, that the Bureau of Agricultural Economics would be given the responsibility for overall planning in the Department of Agriculture, and that the Soil Conservation Service would handle land acquisition, development, and management. The extension agents, working out of the state land-grant colleges, would form state and county land use committees to aid the Department in planning long-range programs for agriculture.

While Wallace moved to reorganize the Department of Agriculture, the Administration also launched a massive propaganda campaign in the farm belt to win support for the new law. The new bill gave the government more power to control production than ever before, but Roosevelt realized that unless farmers accepted the need for production controls his program would fail. Roosevelt had launched a similar campaign in 1933 to win support for the first AAA, but had faced strong opposition from left-wing radicals. In 1938 the President again faced well-organized opposition, but it would be from the right, not the left.

In the spring of 1938 the first statewide organization of the Corn Belt Liberty League was created at a mass meeting of 3,500 farmers at McComb, Illinois. With major support from the conservative Chicago *Tribune*, the Liberty League soon spread into Illinois,

One of the most commonly used and affordable tractors was the Fordson, shown on this Middle Western farm. Rubber tires, sometimes filled with water to provide more traction, began to replace metal wheels on tractors during the decade. (Nebraska State Historical Society)

Nebraska, Kansas, Missouri, and Iowa. The Farmers' Independence Council of America, headed by Dan Casement of Kansas, joined in the attack on Roosevelt. Both organizations insisted that Wallace and Roosevelt were leading the nation down the path of destruction and demanded that the government dismantle its crop control programs. Appealing to the farmers' sense of rugged individualism, they warned that if Roosevelt did not stop violating the natural laws of economics, the result would be the death of freedom.

The new farm bill inevitably became involved in the 1938 election campaign. In a major Republican farm address in Topeka, Senator Arthur Vandenberg, who had emerged as one of the New Deal's most articulate opponents, denounced the Roosevelt Administration's handling of the farm problem and promised that if enough Republicans were returned to Congress in 1938 they would support the McNary-Haugen two-price system for agriculture. Senator Capper, although he had voted for the Agricultural Adjustment Act of 1938, joined in the Republican attack and indicated that he was almost convinced that the only way out of the depression was to return to unrestricted production and competition in agriculture. Other supporters of a two-price system for agriculture tried to persuade Roosevelt and Wallace to reevaluate previous cost-of-production plans.

Congressman Edward Eicher of Iowa introduced a bill in Congress which would have allowed farmers unlimited production but would have assured them at least cost-of-production prices, with a guaranteed profit, on the portion of their crops that was consumed domestically. With wheat selling at fifty cents a bushel and corn for only forty cents a bushel, both plans gained widespread support in the Middle West.

As unrest mounted, Republicans were optimistic that Roosevelt's farm program would cause farmers in the Middle West to turn their backs on the New Deal. Republican spokesmen, led by John D. Hamilton, chairman of the Republican National Committee, charged repeatedly that the AAA would lead to regimentation, that it failed to resolve the paradox of glutted surpluses while millions of people were still hungry, that the reciprocal trade agreements program allowed cheap agricultural products to flood into the country, and, most important, that the New Deal had failed to raise agricultural prices.

During the summer and fall Secretary Wallace campaigned hard and traveled extensively throughout the Middle West defending the New Deal's farm program. Wallace insisted that the farm program was sound and had not been given a fair trial since it had been passed too late to control the production of many crops, including the winter wheat crop in the Middle West. Wallace was also quick to use the discretionary features of the new legislation to bring immediate cash benefits to disgruntled farmers. In July Wallace announced that wheat farmers would receive parity payments from the government to compensate for low market prices. The Secretary also promised that the government would try to raise prices by subsidizing the export of at least 100 million bushels of wheat. During one of his stops, at Hutchinson, Kansas, Dan Casement, head of the Farmers' Independence Council of America, unexpectedly rose from the audience and accused Wallace of trying to buy farm support. Casement chided Wallace, "You cannot have a planned economy in a democracy, and the AAA is a planned economy which, if carried to its logical conclusion, means the destruction of democracy. If the AAA benefits were discontinued, the AAA would blow up over night." Philosophically many farmers may have agreed with Casement, but Wallace realized that he was hardly running a political risk by making farmers an offer they could not afford to refuse.

Caterpillar tractor with a 21-foot disk, covering 80 to 90 acres a day, on a North Dakota farm, 1937. (State Historical Society of North Dakota)

Wallace also announced a new plan, which would go into effect after the election, to respond to the charge that the New Deal, in spite of overproduction in agriculture, had failed to eliminate hunger and malnutrition. On October 12 he announced that the Department of Agriculture was working on a plan to distribute food to low-income families as well as families who were on relief. While Republican hopefuls promised that the McNary-Haugen plan would enable farmers to dump their surplus production abroad, Wallace offered what he hoped would be a more alluring promise, that farmers could dump their products at home while simultaneously feeding the nation's poor.

In November, in spite of Wallace's efforts, jubilant Republicans handed the New Deal one of its worst defeats, especially in the Middle West. Of the Senate seats in the Middle West that were contested in 1938, the Republicans won five of nine contests, increasing their number in the Senate to eight, while the number of Democrats fell to only twelve. In the House before the elections there were eighty-nine Democrats, thirty-five Republicans, seven Progressives, five Farmer-Laborites, and one vacant seat. After the election the number of Democrats dropped to fifty-four, the number of Republicans jumped to eighty, while only two Progressives and one Farmer Laborite were returned to Congress.

Following the election, reports filtered into the Democratic National Committee that farmer dissatisfaction with continuing mortgage foreclosures, small acreage allotments under the AAA's new crop control program, the issue of compulsory controls, and the failure of the 1938 farm program to sustain farm prices and farm income had hurt Democratic candidates in the Middle West. It was equally apparent, however, that there were too many complex issues involved in the 1938 elections to conclude that the Republican vote represented an absolute rejection of the New Deal's farm program. Roosevelt's efforts to pack the Supreme Court, his purge of the Democratic Party during the 1938 primaries, mounting federal debts, the reciprocal trade agreements program, expensive federal relief programs, and the executive reorganization bill may have hurt Democratic candidates even more.

Wallace was discouraged but refused to believe that the New Deal farm program was a failure or that the vote was a repudiation of the Administration's crop control program. He wrote to William Allen

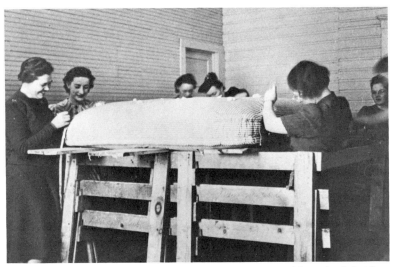

Federal agencies such as the FERA and the FSCC tried to provide work relief while disposing of surplus agricultural products at the same time. Here some Bottineau County, North Dakota, women work on a mattress-making project. (State Historical Society of North Dakota)

White, "We have laid a splendid groundwork and I am confident that the farm program is going to survive in its main essentials no matter how bitter the attacks may be. No our efforts are not futile. We are making progress." If the AAA had made progress, only an incorrigible optimist could have believed that the New Deal had found a solution to the farm problem. Farm prices in 1938 averaged only 78 percent of parity, and gross farm income fell, for the first time since Roosevelt took office, to only $9.3 billion, compared with more than $10 billion in 1937. Farmers who participated in the AAA program received conservation payments totaling $443,691,000 from the government, for complying with the new acreage allotments and for participating in the Administration's soil-conservation program. The number of farmers who participated in the AAA program in 1938 increased by 40 percent over the 1937 figure, but most of the increases occurred among small farmers in the South, and as a consequence only 72 percent of the nation's cropland was covered by the AAA's crop control programs during the year. The number of acres planted to soil-depleting crops was twelve million acres less than the previous ten-year average, but farm production was still 4 percent higher than the 1924-1929 average.

Caterpillar diesel RD4 tractor with a belt pulley threshing barley, yielding 50 bushels an acre, on an 800-acre farm in North Dakota. (State Historical Society of North Dakota)

In the Middle West, low farm prices and surplus production continued to threaten the Administration's efforts to stabilize the agricultural economy. In the corn belt only 93,257,000 acres of corn had been planted during the year, well within the AAA's national goal of from 94 to 97 million acres, but the corn crop, primarily because of the increased use of hybrid seed, still totaled 2,542 million bushels, 160 million bushels more than the average yield. Farmers who participated in the AAA program were paid ten cents a bushel for staying within their acreage allotments and received a total of $61,044,000 in conservation payments during the year. Farmers in the winter wheat belt had already planted their crops before the Agricultural Adjustment Act of 1938 became law. Had the Congress acted more quickly the AAA had planned to establish a national allotment for wheat of only 62,500,000 acres. The actual acreage planted to wheat in 1938 was nearly 17 million acres more than the AAA's goal and produced a crop of 1,084 million bushels, 374 million more than what was needed to meet the foreign and domestic demand. Farmers in the winter wheat regions were paid $1.25 an acre, under the old AAA program, for diverting their acreage from wheat production and received a total of $127,642,000 in conservation payments in 1938. In the spring wheat acres farmers who planted within their acreage allotments were paid twelve cents a bushel for limiting their production during the year.

To keep farm prices from declining even further, the Commodity Credit Corporation offered loans to both corn and wheat producers in 1938. As the year began, the CCC already held 46 million bushels of corn which had been sealed under the 1937 program. To keep the surplus from flooding the market, the CCC offered farmers a new loan, at the rate of 57 cents a bushel, to keep their corn in storage for another year. Many farmers took advantage of the higher loan rate, and eventually 30 million bushels of corn from the 1937 crop were resealed. Under the 1938 loan program additional loans, at the same rate of 57 cents a bushel, were offered to farmers, and the CCC placed another 227 million bushels of corn in storage. The CCC also faced the problem of trying to manage the gigantic wheat crop. Some of the pressure created by the wheat surplus was relieved by wheat exports, which totaled 107 million bushels during the year, 94 million of which were exported with the aid of government subsidies. Still, under the wheat loan program, which lasted from December 1938

By the end of the 1930s many farmers had learned to control wind erosion through disc terracing. Improved soil conservation methods did nothing however, to limit agricultural production. (Kansas State Historical Society)

until March 1939, the CCC made loans, at the rate of 53 cents a bushel, on 85,745,000 bushels of wheat. By the end of the year the threat that the nation's granary, rather than being ever-normal, would become ever-abnormal had already become a reality.

Dissatisfaction with the New Deal's farm program continued in 1939. Even the AFBF, while it still insisted that the New Deal's approach to agriculture was sound, joined in the attack against efforts to expand the New Deal's welfare state to include other dis-advantaged groups in society. Farmers, complaining that WPA wages were too high, that federal relief agencies were inefficient and wasteful, and that further centralization of authority in Washington undermined the free enterprise system, vigorously opposed new ideas and programs for urban Americans that required additional federal spending. The AAA also had few supporters in the city. The Roosevelt Administration's argument that an increase in farm buying power would automatically benefit industrial workers by providing a general stimulus to the economy, was openly questioned by American labor. Although the New Deal had spent millions of dollars trying to raise farm income, industrial workers still faced long unemployment lines, and the economy, mired in recession, appeared to suffer from permanent stagnation.

Ironically, the Farmers' Union, which in earlier years had been considered a ''left-wing'' opponent of the New Deal, was one of the few consistent supporters of the Roosevelt Administration in the farm belt. The Farmers' Union, unlike the Grange, the Farmers' Liberty League, and the Chamber of Commerce, called for more, not less, planning to restore prosperity to agriculture. When William Lemke and Lynn Frazier introduced a cost-of-production bill in the Congress which would have eliminated production controls, the Farmers' Union and a group of grain cooperative spokesmen, headed by Myron W. Thatcher and Glenn Talbott, demanded that the bill be defeated. Dumping American farm products abroad was emotionally appealing, but, even if possible, it would have been irresponsible. Production controls remained unpopular, but proposals that would have allowed unlimited production were worse than no solution at all.

Secretary Wallace continued to try to build support for the Administration's farm program. With the threat of war hanging over Europe, Wallace insisted that a healthy agriculture was necessary to challenge the ''dictator'' nations of the world. Wallace conceded that

Secretary Hull's reciprocal trade program had not brought back more than a small portion of the farmers' lost markets, but he insisted that "the actions of dictators have kept that policy from meeting full success." At the annual Jefferson Day Dinner Wallace appealed to Congress, on patriotic grounds, to support the Administration. He concluded, with impassioned rhetoric, "The national farm program is the American farmers' answer to the thunderings of dictators. This program is Jeffersonian democracy adapted to the needs of today."

Wallace's scapegoating was hardly necessary to guarantee that Congress would continue to fund the New Deal's farm program. Conservatives complained about the cost, but it was no longer possible to question the wisdom, or the practicality, of subsidizing the nation's farmers. The farm appropriation bill debated during the spring of 1939, totaling $1.2 billion, was the most expensive in the nation's history, increasing the cost of the previous farm bill by nearly $400 million. The bill included $225 million for parity payments and another $113 million for the removal of surplus agricultural products. In spite of a lot of rhetoric about balancing the budget, Middle Western Congressmen overwhelmingly voted for the appropriation. Urban Congressmen also agreed to support the bill, provided the farm bloc would in turn promise to vote to fund their programs. Roosevelt, who was fighting to reduce federal expenditures, pressured the farm bloc to cut back their demands. With the 1940 elections on the horizon, however, he too capitulated and in June signed the appropriation bill into law.

Congress had again endorsed the New Deal's approach to the farm crisis. Farmers, and their legislators, rushed to share in the federal largesse, not so much because they believed in Roosevelt and the New Deal, but rather because they could not resist the siren call of the federal treasury. The farm community's reliance on the government appeared to be complete and irreversible.

Chapter 9

Santa Claus Hasn't Been Shot Yet

By the end of the 1930s a number of agrarian spokesmen, including Francis David Farrell, the president of Kansas State University, warned farmers that if the trend toward mechanization — and its handmaiden, increasingly large economic units — continued, farming would become nothing more than a business enterprise that would destroy the comparatively simple and independent lifestyle of the American farmer. Farrell urged a return to subsistence farming

Workers string wire to bring electricity to a Middle Western farm.

Electric motors, which were used to pump water, saw wood, shell corn, mix feeds, and run milk separators, were valuable assets that dramatically increased a farmer's efficiency. This workshop shows the varied uses of electricity on the farm. (South Dakota State Historical Society)

techniques, but he found that few farmers wanted to turn back to the isolation and relative poverty of the past. Their choice symbolized the beginning of a new age in American agriculture and the end of an agrarian dream.

The commercialization of agriculture was a long-range development deeply rooted in the nation's history. Roosevelt's opponents, however, tried to place most of the blame for the increase in tenancy and the growth of large economic units, especially in the Middle West, on the New Deal's farm policy. Kansas Senator Arthur Capper and Congressman Ed Rees, also from Kansas, charged that the AAA discriminated against small farmers while "suitcase" farmers and absentee landlords received huge benefit payments from the government. Rees pointed out that nearly a million farmers who participated in the programs received less than $20 each for cooperating with the AAA, another 750,000 received from $20 to $30 each, 500,000 received only $40 to $50 each, and another 500,000 from $50 to $100 each. The top 20 percent of the nation's farmers received nearly two-thirds of the benefit payments distributed by the AAA, leading

Rees to the obvious conclusion that America's agricultural ''elite'' had benefited most from the New Deal.

Capper charged that government credit policies were driving farmers from the land. The Kansas Senator accused the government, which as a result of Roosevelt's extensive agricultural credit program was now the nation's largest real estate operator, of foreclosing mortgages on sixty thousand farms, most of which were in the Middle West. Capper, although he frequently complained about the ''profligate'' spending of the Roosevelt Administration, now urged the President to abandon ''ordinary prudent business policies'' to save the small family farmer. To keep farmers on the land, he urged the government to lower interest rates to just 3 percent and declare a moratorium on principle payments until the summer of 1943.

There was also a growing awareness that the farm tenancy program, which had been hailed with such enthusiasm in the Middle West, was nothing more than a ''teaser.'' In 1939 the number of

Electric lights replaced kerosene lamps in many Middle Western homes and schools. Electricity brought smiles to these young school children, but it also increased the farmers' efficiency and reduced the number of people needed to work the land. (State Historical Society of North Dakota)

Electricity, made possible by the REA, dramatically improved the quality of life on many Middle Western farms in the 1930s. Electric lights and appliances made life easier and became important symbols of modernization. (South Dakota State Historical Society)

applications under the tenant land purchase program outnumbered the actual loans granted by thirty-four to one; only 4,340 loans, averaging $5,602, were granted during the year. In the Middle West, by the fall of 1939, only 968 loans, totaling $7,779,491, had been granted under the program. Congressmen from the Middle West attacked the New Deal for failing to solve the tenant problem, but they also realized that the economy-minded Congress would not tolerate another ''raid'' on the treasury. Congressman Clifford Hope accurately observed, ''Santa Claus hasn't been shot yet, but there is, of course, a definite conclusion that sooner or later expenditures must be reduced and economy put into effect. This of course militates against any program which calls for increasing indebtedness.''

Government spokesmen like Secretary of Agriculture Wallace faced the impossible task of defending the virtues of the old agrarian order while simultaneously pushing for modernization and change. Emotionally Wallace wished to save small farmers. Intellectually he believed that there were already too many farmers in the United States, and that larger economic units were the inevitable wave of the

The number of telephones on farms actually declined in many areas as farm families sought ways to make ends meet. Still, the telephone broke down rural isolation and became another major factor in modernizing American agriculture. This crew is putting up a telephone pole. (South Dakota State Historical Society)

With the coming of electricity, many farmers bought their first radios. The radio provided news and entertainment, and linked farmers to an increasingly complex world. (State Historical Society of North Dakota)

future. Like his contemporaries he believed that technology, planning, and efficiency would bring prosperity to the farm. He simply could not turn his back on "progress." The result was that at best Wallace appeared to be ambivalent; at worst he seemed unctuous and inconsistent. He admitted that large-scale farmers might have used government payments to buy machinery and to increase the size of their landholdings, but he also argued that even small government checks had enabled many small farmers to remain on the land. Wallace was undoubtedly right. But he also understood that small farmers could not compete in the marketplace with larger farmers and would live in relative poverty unless they became more productive. More efficiency and productivity, however, caused overproduction — which would also drive them from the land. It appeared they could not win.

Spike Evans, who served as administrator of the AAA in the late 1930s, also defended the AAA against the criticism that it was, in effect, creating a heavily subsidized landed gentry in the United States. He conceded that most AAA participants received relatively small benefit payments but was quick to add that of the total of 8,850,914 farmers who participated in the 1936 program, only 12,746 had received more than $1,000, only 385 more than $5,000, and just 91 more than $10,000. In the Middle West the size of farms varied dramatically, ranging from fewer than 120 acres in eastern Kansas to more than 2,000 acres in the Sandhills of Nebraska, but small farms of 300 to 400 acres still dominated the Middle Western landscape.

AAA payments did benefit large landowners more than small farmers. Since entire farms could be purchased for $3,000 to $5,000, many farmers used their payments to invest in more land. In the early years of the depression the size of many farms in the Middle West actually decreased, but after 1935 the size of farms increased substantially. In Kansas the average farm in 1930 was 283 acres; by 1940 the average size was 308 acres. Nebraska went from an average of 345 acres in 1930 to 391 acres in 1940. South Dakota jumped from 439 acres in 1930 to 545 acres in 1940. Government payments by themselves, however, did not cause large farming units to appear. The trend toward mechanization and larger units had been apparent since the nineteenth century, long before the New Deal came on the scene. By the late 1930s nearly 90 percent of the commercial farm

Farmer working his fields with a team of mules. (State Historical Society of North Dakota)

production in the United States came from only 50 percent of the nation's farmers. Some corporations that had invested heavily in land did receive huge benefit payments from the AAA. In 1937, for example, the Metropolitan Insurance Company received $257,095 from the AAA. In 1939, in a move that was intended to satisfy New Deal critics, AAA payments were limited to a total of $10,000. Corporate farms, however, played a relatively minor role in struc-

For many Middle Western farmers, the 1930s was a period of transition from man and horse power to machine power. Here a tractor pulls a hand-held plow. (Nebraska State Historical Society)

turing the farm economy in the Middle West. Limiting the size of benefit payments to large farmers, unfortunately, did nothing to improve the plight of the nation's small family farmers.

The Department of Agriculture and the Bureau of Agricultural Economics also released statistics which appeared, on the surface, to support the AAA's contention that it was not driving farmers off the land. The total farm population, which in 1938 experienced the largest increase of any year since 1932, was 32,059,000 on January 1, 1939, compared with 31,819,000 on January 2, 1938, and the all-time record high of 32,077,000 on January 1, 1910. The population increase, however, obscured the fact that the number of individual farming units and the number of people who were actually engaged in farming was declining. In 1938, for example, it was estimated that 828,000 people had moved to farms while another 1,025,000 had left their farms. The net loss of more than 200,000 people was, however, offset by the surplus of births, which in 1938 totaled 747,000, over rural deaths, which totaled only 305,000. The Great Plains area was especially hard hit by the out-migration, losing as many as 300,000 people during the decade. In many rural counties the number of farms declined by more than 25 percent. The population of states like Minnesota, Iowa, and Missouri was more stable but still declined by an average of 7 percent during the decade.

Thousands of dams, reservoirs, and farm ponds were constructed during the 1930s. This photograph shows a completed detention dam in Brown County, Kansas. (Kansas State Historical Society)

The governor of the Farm Credit Administration, F. F. Hill, also defended the FCA against charges that it had failed to save small family farmers. Hill insisted that the FCA's policies, while prudent, had greatly benefited the nation's farmers. After the drought of 1936 the FCA had gradually scaled down its emergency and long-term loans, but it continued to be the major source of credit for most farmers. In the Middle West from 1937 to 1939 the FCA made 136,740 Emergency Crop and Feed Loans, totaling $19,877,525, and 40,521 Federal Land Bank Commissioner Loans, totaling $125,639,600. The government did foreclose on thousands of mortgages in the late 1930s. To limit government losses, the land was then sold for whatever price the market would bear. Hill proudly pointed out, however, that the rate of mortgage foreclosures, which had reached a peak of 38.8 per 1,000 in 1932, had fallen to only 13.5 per 1,000 in 1938, the lowest rate since 1926. The FCA undoubtedly saved many farmers, but as many as 1,500,000 farmers lost their farms through mortgage foreclosures between 1930 and 1939. In many Middle Western communities, one in ten farms was foreclosed during the decade. Farm indebtedness was the lowest in twenty years, but many farmers were still heavily in debt. In North Dakota, for example, it was estimated that more than 75 percent of the farms were still mortgaged, that taxes were delinquent on 70 percent of the farms, and that perhaps as many as 35 percent of the people in the state were still receiving federal relief. More liberal credit policies might have kept more farmers on the land in the 1930s, but it is more likely that cheaper credit would have accelerated mechanization, which in turn would have reduced the number of people needed to work the land.

Urbanization continued in the 1930s, but the lack of economic options in the city, combined with government relief programs, probably slowed the process. Several million people who left the farm in the 1920s returned when they lost their jobs in the city. As soon as they had a chance, they again left the farm. Between 1935 and 1945 the number of farms in the United States declined by nearly 1.7 million. The flight from the land accelerated even more dramatically after the Second World War. By 1980 the nation's farm population had fallen to only 6,051,000 and the total number of farms to only 2,248,000.

Poor farmers who hung onto their farms in the 1930s continued to

Workers complete a grade stabilization structure on this Middle Western farm. (Kansas State Historical Society)

receive assistance from the FSA. By the end of 1939 some 1,231,000 farmers throughout the country had been aided by the FSA. Their status, however, had changed very little. A survey in 1939 of 360,015 families who had been aided by the FSA measured the government's success, and failure. The survey indicated that the FSA's clients had increased the size of their acreage, enjoyed better diets, increased the number of their farm animals, were more self-sufficient, and, if tenants, had generally secured more favorable leases from their landlords. The survey also indicated that the income of the families averaged only $375 when they joined the program, but averaged $538, a rise of 43 percent, by 1939. The problem was not that the FSA was not making progress, but that its clients had so far to go to escape their poverty.

The plight of migratory-casual workers was even worse. In November 1939 the FSA issued a report that indicated that as many as 350,000 migratory farm families had annual incomes that ranged from only $200 to $450 a year. Many of the families had lost their farms through foreclosure during the early days of the depression; others had been forced from agriculture by drought; even more were "tractored" out when they were replaced by machinery. By 1939 there were twenty-six permanent migrant camps in operation or under construction which could care for seven thousand families at one time. John Steinbeck's novel *The Grapes of Wrath* created sympathy for the migrants, but there seemed to be no place for them in the modern industrial state.

Migrant workers were among the first of the nation's farmers to experience the brutalizing impact of technology. More than 60 percent of the nation's farmers still used horses and mules to work their farms, but the revolution in technology which would transform American agriculture was already well under way. Between 1900 and 1930 the number of man-hours needed for some crops was cut by nearly 50 percent. The number of tractors, trucks, and harvesting machines used on farms, in spite of the depression, increased dramatically during the decade. In the Middle West the number of combines doubled, the number of tractors increased by nearly 60 percent, while the number of cornpickers increased six-fold. For the nation as a whole, there were 246,000 tractors on farms in 1910; by 1939 there were 1,600,000. The Northern Plains was one of the most heavily mechanized regions of the country. In North Dakota, for

Contour farming, shown in this 1939 photograph, proved to be an effective soil conservation technique and was widely practiced by the end of the depression decade. (Kansas State Historical Society)

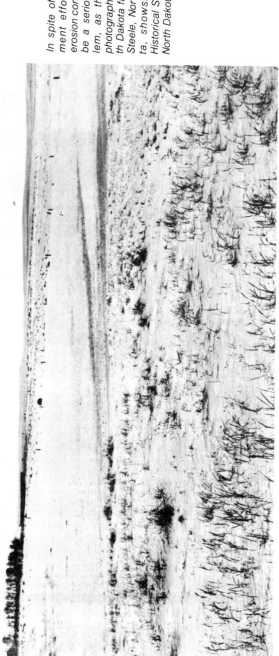

In spite of government efforts, soil erosion continued to be a serious problem, as this 1939 photograph of a North Dakota farm near Steele, North Dakota, shows. (State Historical Society of North Dakota)

example, 57 percent of the farmers had tractors in 1920; by 1930 the number of farmers with tractors had increased to 87 percent. The one-way disc plow, which cut a ten-foot swath, had been introduced earlier in the century, allowing farmers to plant more acreage faster than ever before. The Rural Electrification Administration, created in 1935, not only made life on the farm more pleasant, it also made farmers more efficient and productive. Most farms still had no electricity when World War II began. In North Dakota, only 2 percent of the farmers were electrified in 1939; Nebraska had the highest number of farms with electricity, 28 percent. Still, many farmers now had access to an unlimited supply of cheap power. The quality and types of machinery used varied widely, but by some estimates farm labor became more than 30 percent more efficient during the decade. The use of hybrid seeds, pesticides, fertilizers, and irrigation was still in its infancy but had already begun to have a major impact on productivity. Wheat production per acre increased only slightly, but hybrid corn yields doubled in many areas of the Middle West. The government still tried to control production by limiting acreage, but farmers only produced more on less land. Each year the number of people needed in agriculture to feed the population declined. In 1930 it took one farmer to feed 9.8 people in the United States; by World War II one farmer could feed 14.6 people.

Farmers could create abundance, but the Roosevelt Administration had still not resolved the embarrassing paradox of want in the midst of plenty. Millions of Americans were malnourished and still lived at near-starvation levels. Relief officials estimated in 1939 that forty million Americans were living in families where the average cash income was only $9.00 a week, and that they spent, individually, only about $1.00 a week on food, or about five cents a meal. The Federal Surplus Commodities Corporation continued to purchase surplus agricultural products to be distributed to the needy through state and local relief agencies, but its program was limited in scope and had been frequently criticized for interfering with the free enterprise system.

On March 13, 1939, Secretary Wallace announced a new program, the so-called Food Stamp Plan, which he promised would not only feed the hungry and malnourished but would also benefit farmers. The plan proposed that eligible families, normally those already on relief, would buy orange stamps from the government in an amount

This 1938 dust storm appears to threaten all forms of life as it bears down on a small community in the Northern Plains. (State Historical Society of North Dakota)

Skeptics questioned whether trees could survive the heat and dust, but by the end of the 1930s it was apparent that the shelterbelt project would be one of the New Deal's most visible legacies in the Middle West. (Kansas State Historical Society)

equal to their normal food purchases and then would be given free a number of blue stamps, usually one for every two orange stamps they purchased. The orange stamps could be used to purchase any kind of food the buyer wanted; the blue stamps could be used to purchase only surplus agricultural products, especially fruit and vegetables, which had been identified in advance by the government. Grocery stores that received the stamps would then turn them back to the government for payment in cash. It was hoped that the poor would consume more and better food and that a "new" market for American farm products would be tapped for the first time. Since the plan worked through the existing system of distribution, it won immediate approval from the farm community and from the nation's food dealers.

The plan, which was first implemented on an experimental basis in Rochester, New York, and Dayton, Ohio, in the spring of 1939, would fall far short of its initial promise. Although the program had been extended to nearly half of the country's nearly three thousand counties within four years, only $261 million had been spent to finance the Food Stamp Plan before it was canceled in 1943. An eventual casualty of the war, the plan had been developed too late to have a major impact on the depression-stricken economy of the

In spite of a massive government effort to promote soil conservation, soil erosion continued to be a serious and unresolved problem at the decade's end. (Kansas State Historical Society)

1930s. Had it been tried earlier it would still have been only a prop, not a solution to the combined problems of surplus agricultural production and needless privation. The Food Stamp Plan could have eliminated much of the hunger and malnutrition that existed during the depression, but only profound changes in the structure of the economy, to give the poor a more balanced share of the nation's wealth, would have enabled the impoverished masses to escape the poverty of abundance.

As the New Deal drew to a close it was called upon again to do what it did best, to provide emergency drought relief. In the Middle West only Kansas, Nebraska, and North and South Dakota suffered from the full effects of drought in 1939, but many farmers in other states also suffered from dry weather. During the summer and fall, in what was by now a well-rehearsed routine, the Farm Security Administration began making emergency grants to drought-stricken farmers, while other New Deal agencies, such as the Works Progress Administration, the Civilian Conservation Corps, and the Federal Surplus Commodities Corporation, also provided immediate relief. By November the FSA was providing emergency grants to nearly nineteen thousand farm families in North Dakota, South Dakota, and Nebraska. Again, federal drought relief was a godsend. Senator Arthur Capper reflected the feelings of many farmers when he wrote, "I have differed with President Roosevelt on many of his policies, especially on foreign relations and on his extravagant spending and lending programs — but I will say for him that he never has turned a deaf ear to the relief needs of distressed farmers."

Drought continued to focus attention on the Administration's soil-conservation programs. By the end of the 1930s a number of local, state, and federal agencies were working to promote soil conservation. In many areas of the Middle West farming practices had changed. More land was planted to sorghum and sudan grass, more acreage was left fallow, terraces and farm ponds had been constructed, new plowing techniques helped to control wind and water erosion. The Great Plains Shelterbelt Project continued; by the end of the 1930s some 220 million trees, with a survival rate of nearly 80 percent, had been cultivated. Advances had been made, but a study by the Soil Conservation Service indicated that as many as two hundred million acres of land were still subject to moderate to severe erosion and estimated that only 75,000 farm families, out of

Wind erosion threatened the nation's most vital resource, the soil, throughout the depression decade. (The Land and Its People Museum, Red Cloud, Nebraska)

3,600,000, were working fully protected land. By 1939 the government had purchased more than nine million acres of submarginal land, but it was still fighting an uphill battle. In 1940 the Great Plains Committee concluded in its final report, *The Northern Great Plains*, "There is comparatively little to show in the way of long-term rehabilitation for the huge sums spent there by the Federal Government in recent years." The report continued, "The problem of land-use adjustment on an enduring basis in the Great Plains, in the Northern Plains and Southern Plains alike, still remains the most difficult agricultural problem of its kind in the United States."

Overproduction and low farm prices still plagued farms on the eve of the Second World War. In 1939, 5,756,240 farmers signed up for the AAA program, an increase of 10 percent over 1938 and 57 percent over 1937, but still only 78 percent of the cropland in the United States was covered by AAA contracts. Cattle and dairy products were selling at near parity levels, but the price of corn was only 59 percent of parity, wheat 50 percent, and hogs only 60 percent.

In 1939 the government set the wheat allotment at fifty-five million acres, but only 73 percent of the nation's wheatgrowers participated. In spite of drought, another bountiful harvest, totaling 755 million bushels, glutted the market. To bolster the price of wheat, farmers who had signed AAA contracts were given conservation payments of 17 cents a bushel, an 11-cent-per-bushel parity payment, and wheat loans, from the Commodity Credit Corporation, averaging 63.3 cents a bushel. As a consequence of the government's price support program, wheat farmers who had joined with the AAA were guaranteed at least 91 cents a bushel for their 1939 crop. Wheat still sold for only 54 cents a bushel on the open market, but wheat farmers increased their income in 1939 to $534,267,000, compared with $446,500,000 in 1938 and only $199,800,000 in 1932.

Government efforts to raise the price of corn by limiting production also failed. The problem, however, was not, as in the wheat growing areas, noncompliance, but rather the failure of the administration to anticipate the increase in acreage yields due to the more extensive use of hybrid seed corn and the rapid mechanization of the corn-hog industry. In 1939 the AAA had established the goal of limiting corn acreage to between ninety-four and ninety-seven million acres. Although only ninety-one million acres were actually

Drought and grasshoppers persisted as the 1930s came to an end. A field destroyed by grasshoppers, 1939. (State Historical Society of North Dakota)

planted, the harvest totaled 2,619,000 bushels, nearly 300,000 bushels more than the 1928 to 1937 average. The year's production, combined with the carryover from previous years, made 3,202,000,000 bushels available for marketing in the 1939-1940 crop year. With abundant feed supplies, hog production jumped to record levels as seventeen billion pounds of pork flooded the market. In an effort to stabilize prices the government provided farmers who had participated in the AAA program a conservation payment of nine cents a bushel, parity payments of six cents a bushel, and loans from the Commodity Credit Corporation at fifty-seven cents a bushel. Although prices, which averaged fifty cents a bushel during the 1938-1939 marketing year and fifty-six cents a bushel in 1939-1940, were kept at near-parity levels by government subsidies, the AAA's crop control was in serious financial trouble before World War II came to the rescue.

Wallace's dream of "progressive balanced" abundance was now replaced by the fear that the Commodity Credit Corporation was on the same road that had bankrupted Hoover's Federal Farm Board. Under the 1938 wheat program the CCC made seventy-three thousand loans, at the rate of fifty-three cents a bushel, totaling $46,430,000. When farm prices failed to rise, many farmers chose not to redeem their loans, saddling the government with a $5,620,000 loss. In 1939, 237,000 loans, on 168 million bushels of wheat at 63.3 cents a bushel, were extended to wheat farmers under the CCC program. By early 1939 the CCC already held 257 million bushels of corn. The government resealed 166 million bushels by offering farmers another loan, but many farmers, who decided not to take advantage of the new loans, gave the CCC the remaining 91 million bushels of stored corn to pay off their loans. By December 1939 loans on the 1939-1940 corn crop were made available, at the rate of seventy-five cents a bushel. Before the program closed, the government had made loans on another 301,909,000 bushels, leaving the CCC with more than 500 million bushels of corn in storage. By the end of the 1930s the wheat reserve was double and the corn reserve triple what it had been when the decade began. The granary was about to burst.

By the end of the year real farm income, which totaled $8,519,000,000, while still below the 1937 level was 72 percent higher than in 1932 and had restored farm buying power to a level

Surplus foods were stored and distributed to the nation's poor throughout the 1930s. The food distribution program was, however, politically unpopular and was never viewed as a solution to the problem of overproduction. (Nebraska State Historical Society)

roughly equal to that of 1929. Although farm income had risen, the farmers' income had become directly dependent upon the continuation of government payments. In the Middle West, farmers received $234,560,366 in conservation payments in 1939 and $81,978,706 in parity payments. Other federal expenditures also brought relief to every section of the Middle West, where, particularly in the drought states, they ranked among the highest in the nation. Between 1933 and 1939 federal agricultural expenditures, combined with agricultural loans, reached $746 per capita in North Dakota, $688 per capita in South Dakota, $677 in Nebraska, and $493 per capita in Kansas.

Still, many farmers in the Middle West viewed the AAA as a temporary emergency measure. New Deal planners still hoped to create a stable, rational, predictable social order in rural America, but many farmers dreamed of returning to a system of laissez-faire economics with unlimited production. Foreign markets, however, remained closed. The embarrassing paradox of want and abundance continued. When war broke out in Europe in 1939, a new era appeared to be born. Another bitter debate began about whether to continue the AAA. Farmers clearly wanted federal subsidies to continue, but a poll of farmers in the Middle West indicated that only 58 percent of those interviewed believed that the AAA should continue to control production. Temporarily, the war had ended the dread of plenty.

A Note on Sources

Space permits only a brief discussion of the major sources used in this study. The papers of Franklin D. Roosevelt, Louis Bean, Morris L. Cooke, Harry Hopkins, Emil Hurja, Gardner Jackson, Lorena Hickock, Henry Morgenthau, Jr., Eleanor Roosevelt, Charles Taussig, Rexford Tugwell, and Aubrey Williams at the Franklin D. Roosevelt Library in Hyde Park, New York, are indispensable. Personal correspondence of agency heads, as well as official government records, are in the National Archives in Washington, D.C. Of particular importance are Record Group 16, Secretary of Agriculture, General Correspondence; Record Group 44, Records of Government Reports, Division of Field Operations, Statistical Section; Record Group 83, Records of the Bureau of Agricultural Economics, Statistical and Historical Research; Record Group 96, Farm Security Administration; and Record Group 145, Agricultural Adjustment Administration. The Manuscript Division of the Library of Congress, in Washington, D.C., houses the papers of Charles McNary, George Norris, Francis B. Sayre, and William Allen White, as well as the scrapbooks of Henry A. Wallace. The Herbert Hoover Library, in West Branch, Iowa, should be consulted for the 1920s and the Hoover period.

The personal papers of figures who played important roles in developing a response to the farm crisis in the Middle West are scattered. Especially important to students of agricultural history are the papers of Arthur Capper, Alfred M. Landon, and George McGill at the Kansas State Historical Society, Topeka; the M. L. Wilson papers at Montana State University, Bozeman; the Charles Bryan, Val Kuska, and Samuel McKelvie collections at the Nebraska State Historical Society, Lincoln. The Special Collections Department at the University of Iowa, Iowa City, includes the papers of Fred

Biermann, Lester Dickinson, Edward Eicher, R. M. Evans, Clyde
Herring, Christian Ramseyer, Dan Wallace, and especially Milo
Reno and Henry A. Wallace; the papers of Royal Copeland and
Arthur Vandenburg are in the Historical Collections Division at the
University of Michigan, Ann Arbor; the Western History Collection
at the University of Oklahoma in Norman houses the papers of John
A. Simpson and Elmer Thomas; finally, the Western Historical
Manuscript Collection at the University of Missouri in Columbia has
papers for Clarence Cannon, Chester Davis, William Hirth, Peter
Norbeck, and George N. Peek. The Columbia Oral History Project at
Columbia University in New York has important interviews with
Paul Appleby, Louis Bean, Mordecai Ezekiel, Gardner Jackson,
Howard Tolley, and especially Edward O'Neal and Rexford
Tugwell.

Most government agencies published systematic reports that detail
their activities in the 1930s. Yearly reports were issued by the
Civilian Conservation Corps, the Farm Credit Administration, the
Rural Resettlement Administration and the Farm Security Administration, the FERA and the WPA, the United States Department of
the Interior, the Soil Conservation Service, and the Agricultural
Adjustment Administration of the United States Department of
Agriculture. *The United States Census for Agriculture*, the annual
Yearbook for Agriculture, and *Agricultural Statistics, 1933-1950*
(1950) provide important statistical information about agriculture
during the decade. Congressional debate on farm legislation can be
found in the *Congressional Record*. Hearings before the Committee
on Agriculture, House of Representatives, and the Senate Committee
on Agriculture and Forestry in 1932, 1933, and 1937 are very useful.

Hundreds of special government reports were issued in the 1930s.
Among the most important for this study are: *The Future of the Great
Plains* (1937); *The Western Range* (1936); *Activities of Federal
Emergency Agencies 1933-1935* (1938); *Farm Tenancy: Report of
the President's Committee* (1937); *The Northern Great Plains*
(1940); E. L. Kirkpatrick, *Analysis of 70,000 Rural Rehabilitation
Families* (1939); Russell Lord, *To Hold This Soil* (1938); E. A.
Schuler, *Social Status and Farm Tenure — Attitudes and Social
Conditions of Corn Belt and Cotton Belt Farmers* (1938); Carl C.
Taylor, Helen W. Wheeler, and E. L. Kirkpatrick, *Disadvantaged
Classes in American Agriculture* (1938); John D. Hoyt, *Drought of*

1936 and *Droughts of 1930-1934* (1936); *Possibilities of Shelterbelt Planting in the Plains Region* (1935); Berta Asch and A. R. Mangus, *Farmers on Relief and Rehabilitation* (1937); Francis D. Cronin and Howard W. Beers, *Areas of Intense Drought Distress, 1930-1936* (1937); *Current Statistics on Relief in Rural and Town Areas* (1936-1939); R. S. Kifer and H. L. Stewart, *Farming Hazards in the Drought Area* (1938); Irene Link, *Relief and Rehabilitation in the Drought Area* (1937); Charles Lively and Conrad Taeuber, *Rural Migration in the United States* (1939); A. R. Mangus, *Changing Aspects of Rural Relief* (1938) and *Rural Regions of the United States* (1940); Conrad Taeuber and Carl O. Taylor, *The People of the Drought States* (1937); Carl C. Wynee, Jr., *Five Years of Rural Relief* (1938); and Carl C. Wynee and Nathan L. Whetten, *Rural Families on Relief* (1938).

Newspapers consulted include the Des Moines *Register*, the *Iowa Farm Register*, the Liberal *News*, the Mason City *Globe Gazette*, the Minneapolis *Tribune*, the *New York Times*, the St. Louis *Post Dispatch*, the St. Paul *Pioneer Press*, the Sioux City *Journal*, the *Southwest Daily Times*, the *Southwest Tribune*, and the Topeka *Daily Capitol*. Two rural newspapers, *Capper's Weekly* and *Wallace's Farmer*, are particularly helpful. Contemporary magazine articles evaluating New Deal farm programs in the Middle West can be found in many popular magazines including the *Literary Digest*, the *Nation*, the *New Republic*, and the *New York Times Magazine*.

Significant books, including memoirs, by contemporaries include: John Morton Blum, *From the Morgenthau Diaries, Years of Crisis, 1928 1938* (1959); James A. Farley, *Jim Farley's Story: The Roosevelt Years* (1964); Herbert Hoover, *Memoirs* (1951-1952); Harry Hopkins, *Spending to Save: The Complete Story of Relief* (1936); Cordell Hull, *The Memoirs of Cordell Hull* (1948); Harold L. Ickes, *The Secret Diary of Harold L. Ickes: The First Thousand Days, 1933-1936* (1954); Raymond Moley, *After Seven Years* (1939); Arthur Mullen, *Western Democrat* (1940); George W. Norris, *Fighting Liberal: An Autobiography of George Norris* (1945); George N. Peek, *Why Quit Our Own* (1936); Frances Perkins, *The Roosevelt I Knew* (1946); Elliot Roosevelt (ed.), *F.D.R.: His Personal Letters, 1928-1945* (1950); Samuel I. Rosenman (comp.), *The Public Papers and Addresses of Franklin D. Roosevelt (1938-1950)*; John A. Simpson, *The Militant Voice of Agriculture* (1934); Lawrence Svobida, *An Empire of Dust* (1940); Rexford Tugwell, *The Brain Trust* (1968); and Henry A. Wallace, *New Frontiers* (1934) and *Democracy Reborn* (1944).

It would be futile here to try to replicate existing bibliographies which cover the New Deal, agriculture, and the Middle Western farm experience during the 1930s. The sources listed below are quite selective. For a comprehensive bibliography of scholarly articles on the New Deal see Robert E. Burke and Richard Lowitt, *The New Era and the New Deal 1920-1940* (1981); for agriculture see also Gary McDean, *A Preliminary List of References for the History of Agriculture during the New Deal Period 1932-1940* (1969), and from the Agricultural History Center, University of California, Davis, *A List of References for the History of Agriculture in the Great Plains* (1976). For an evaluation of recent historiographical trends on New Deal agriculture see Theodore Saloutos, ''New Deal Agricultural Policy, An Evaluation,'' *Journal of American History* 61 (September 1974). See also Allan G. Bogue, ''The Heirs of James C. Malin: A Grassland Historiography,'' *Great Plains Quarterly* (Spring 1981). For a more general discussion of recent books and research relating to agricultural history, see Gilbert C. Fite's published lecture, ''Recent Trends in United States Agricultural History'' (Texas Tech University, 1985).

Biographies of important figures who played a role in shaping the history of agriculture in the 1930s include: Edward Blackorby, *Prairie Rebel: The Public Life of William Lemke* (1963); Wellington Brink, *Big Hugh: Father of Soil Conservation* (1951); Charles Searle, *Minister of Relief: Harry Hopkins and the Depression* (1963); Everest Seymour, *Morgenthau, the New Deal and Silver* (1950); Gilbert C. Fite, *George N. Peek and the Fight for Farm Parity* (1954) and *Peter Norbeck: Prairie Statesman* (1948); Paul Kurzman, *Harry Hopkins and the New Deal* (1974); Russell Lord, *The Wallaces of Iowa* (1947); Richard Lowitt, *George W. Norris: The Triumph of a Progressive, 1933-1944* (1978); Donald R. McCoy, *Landon of Kansas* (1966); George H. Mayer, *The Political Career of Floyd B. Olson* (1951); William D. Rowley, *M. L. Wilson and the Campaign for the Domestic Allotment* (1970); Edward L. Schapsmeier and Frederick H. Schapsmeier, *Henry A. Wallace of Iowa: The Agrarian Years, 1910-1940* (1968); and Bernard Sternsher, *Rexford Tugwell and the New Deal* (1964).

The best recent study of the depression is Robert S. McElvaine, *The Great Depression in America, 1929-1941* (1984). Classic studies of continuing importance include William E. Leuchtenburg, *Franklin D. Roosevelt and the New Deal 1932-1940* (1963), and Arthur Schlesinger, Jr., *The Age of Roosevelt (1957-1960)*. Excellent general surveys of agriculture with an emphasis on the modern period include Gilbert C. Fite, *American Farmers: The New Minority* (1981), and John L. Shover, *First Majority, Last Minority: The Transforming of Rural America* (1976). The most thorough study of agricultural policy during the 1930s, focusing on the AAA, is Theodore Saloutos, *The American Farmer and the New Deal* (1982). John D. Hicks and Saloutos' earlier study, *Agricultural Discontent in the Middle West 1900-1939* (1951) is still useful. Richard Lowitt, *The New Deal and the West* (1984) is an important contribution to the scholarship of the period. Other books that help in understanding various aspects of New Deal farm policy include Richard Kirkendahl, *Social Sciences and Farm Politics in the Age of Roosevelt* (1966); Harvey Ray, *Want in the Midst of Plenty: The Genesis of the Food Stamp Plan* (1942); Paul B. Sears, *Deserts on the March* (1935, 1980); Wayne D. Rasmussen *et al.*, *A Short History of Agricultural Adjustment* (1975); David E. Conrad, *The Forgotten Farmers, The Story of Sharecroppers in the New Deal* (1965); Paul Mertz, *New Deal Policy and Southern Rural Poverty* (1978); Donald Holley, *Uncle Sam's Farmers, the New Deal Communities in the Lower Mississippi Valley* (1975); Louis Cantor, *A Prologue to the Protest Movement, the Missouri Sharecropper Roadside Demonstration of 1939* (1969); D. Clayton Brown, *Electricity for Rural America: The Fight for REA* (1980); Van L. Perkins, *Crisis in Agriculture, the Agricultural Adjustment Administration and the New Deal* (1969); Sidney Baldwin *Poverty and Politics* (1968); and Janet Poppendieck, *Breadlines Knee-Deep in Wheat, Food Assistance in the Great Depression* (1986). Books on farm organizations, and the politics of agriculture, are: Gladys Baker, *The County Agent* (1939); Christina McFadyen Campbell, *The Farm Bureau and the New Deal: A Study of the Making of National Farm Policy, 1933-1940* (1966); John A. Crampton, *The National Farmers Union: Ideology of a Pressure Group* (1965); Gladys Talbott Edwards, *The Farmers Union Triangle* (1941); Charles Hardin, *The Politics of Agriculture: Soil Conservation and the Struggle for Power in Rural America* (1952);

Orville M. Kile, *The Farm Bureau through Three Decades* (1948); Everett Luoma, *The Farmer Takes a Holiday: The Story of the National Farmers Holiday Association and the Farmers Strike of 1932-1933* (1967); and John L. Shover, *Cornbelt Rebellion: The Farmers' Holiday Association* (1965). *Red Harvest: The Communist Party and American Farmers* (1982), by Lowell K. Dyson, includes an excellent discussion of Communist activities in the 1930s.

For recent histories of areas of the Middle West relating to this study see James E. Wright and Sarah Z. Rosenberg, eds., *The Great Plains Experience: Readings in the History of a Region* (1978); Carl F. Kraenzel, *The Great Plains in Transition* (1955); Wilmon H. Droz, *Trees, Prairies, and People* (1977); Howard Ottoson *et al.*, *Land and People in the Northern Plains Transition Area* (1966); and Leslie Hewes, *The Suitcase Farming Frontier* (1973). Recent studies of the dust bowl include Guy Logsdon, *The Dust Bowl and the Migrant* (1972); R. Douglas Hurt, *The Dust Bowl, an Agricultural and Social History* (1981); Paul Bonnifield, *The Dust Bowl, Men, Dirt, and Depression* (1979); Donald Worster, *Dust Bowl, the Southern Plains in the 1930s* (1979); and Walter Stein, *California and the Dust Bowl Migration* (1974). The spring 1986 issue of the *Great Plains Quarterly* is devoted to the dust bowl and includes an excellent historiographical essay on recent scholarship by Harry C. McDean. Complete documentation for the text is available through the publisher.

Index

by Ginger Weir

Michael W. Schuyler has been a Professor of History at Kearney State College, Kearney, Nebraska, since 1980. He received his B.A. (1963) in History and Business from Southwestern College, Winfield, Kansas; and his M.A. (1965) and Ph.D. (1969) in History from the University of Kansas, Lawrence. He has been an Assistant Instructor at the University of Kansas (1964-1967) as well as Visiting Professor of History at Southwest Texas State University (1980-1981).

Dr. Schuyler has also served as Acting Dean, Graduate School (1978-1979); Chairman, Faculty Senate (1981-1982); and Chairman, Department of History (1982-1983, 1984-   ), Kearney State College.

He was contributing author to *The Great Plains Experience: Readings in the History of a Region* (University of Mid-America, 1978) and to *Land Stewardship Project Materials* (St. Paul, MN, 1984), and is the author of both journal articles and professional papers in his field. In 1986 he was elected Non-resident Fellow of the Center for Great Plains Studies, University of Nebraska, Lincoln.

The Dread of Plenty analyzes efforts by the Roosevelt Administration to aid farmers battling the combined effects of economic depression, drought, and savage dust storms during the 1930s.

Schuyler begins with a discussion of the origins of the New Deal farm program and the rise and fall of agrarian radicalism during that time. He emphasizes that the great droughts, especially in 1934 and 1936, had a major impact on the development of the New Deal farm program as well as on the evolution of agrarian thought during the decade. He also emphasizes, however, the more enduring threat from technological changes in agriculture — overproduction, and the grim irony of want in the midst of plenty.